THE SCHOOL MATHEMATICS PROJECT

When the SMP was founded in 1961, its main objective was to devise radically new secondary-school mathematics courses (and corresponding GCE and CSE syllabuses) to reflect, more adequately than did the traditional syllabuses, the up-to-date nature and usages of mathematics.

This objective has now been realized. SMP *Books 1–5* form a five-year course to the O-level examination 'SMP Mathematics', *Books 3T, 4* and *5* give a three-year course to the same O-level examination (the earlier *Books T* and *T4* being now regarded as obsolete). *Advanced Mathematics Books 1–4* cover the syllabus for the A-level examination 'SMP Mathematics' and five shorter texts cover the material of the various sections of the A-level examination 'SMP Further Mathematics'. Revisions of the first two books of *Advanced Mathematics* are available as *Revised Advanced Mathematics Books 1* and *2*. There are two books for 'SMP Additional Mathematics' at O-level. All the SMP GCE examinations are available to schools through any of the Examining Boards.

Books A–H, originally designed for non-GCE streams, cover broadly the same development of mathematics as do the first few books of the O-level series. Most CSE Boards offer appropriate examinations. In practice, this series is being used very widely across all streams of comprehensive schools, and its first seven books, followed by *Books X, Y* and *Z*, provide a course leading to the SMP O-level examination. An alternative treatment of the material in SMP *Books A, B, C* and *D* is available as SMP *Cards I* and *II*.

Teachers' Guides accompany all series of books.

The SMP has produced many other texts, and teachers are encouraged to obtain each year from the Cambridge University Press, Bentley House, 200 Euston Road, London NW1 2DB, the full list of SMP books currently available. In the same way, help and advice may always be sought by teachers from the Director at the SMP Office, Westfield College, Hampstead, London NW3 7ST, from which may also be obtained the annual Reports, details of forthcoming in-service training courses and so on.

The completion of this first ten years of work forms a firm base on which the SMP will continue to develop its research into the mathematical curriculum and is described in detail in Bryan Thwaites's *SMP: The First Ten Years*. The team of SMP writers, numbering some forty school and university mathematicians, is continually evaluating old work and preparing for new. But at the same time, the effectiveness of the SMP's future work will depend, as it always has done, on obtaining reactions from a wide variety of teachers – and also from pupils – actively concerned in the class-room. Readers of the texts can therefore send their comments to the SMP in the knowledge that they will be warmly welcomed.   1975

ACKNOWLEDGEMENTS

The principal authors, on whose contributions the S.M.P. texts are largely based, are named in the annual Reports. Many other authors have also provided original material, and still more have been directly involved in the revision of draft versions of chapters and books. The Project gratefully acknowledges the contributions which they and their schools have made.

This book—*Book E*—has been written by

    Joyce Harris   C. Richards
    D. A. Hobbs   R. W. Strong
    K. Lewis   Thelma Wilson

and edited by Elizabeth Smith.

The Project owes much to its Secretaries, Miss Jacqueline Sinfield and Mrs Jennifer Whittaker for their willing assistance and careful typing in connection with this book.

We would especially thank Dr J. V. Armitage for the advice he has given on the fundamental mathematics of the course.

The drawings at the chapter openings in this book are by Penny Wager.

We are much indebted to the Cambridge University Press for their cooperation and help at all times in the preparation of this book.

THE SCHOOL MATHEMATICS PROJECT

BOOK E

CAMBRIDGE UNIVERSITY PRESS
Cambridge
London · New York · Melbourne

Published by the Syndics of the Cambridge University Press
The Pitt Building, Trumpington Street, Cambridge CB2 1RP
Bentley House, 200 Euston Road, London NW1 2DB
32 East 57th Street, New York, NY 10022, USA
296 Beaconsfield Parade, Middle Park, Melbourne 3206, Australia

© Cambridge University Press 1970

Library of Congress catalogue card number: 68-21339

ISBN: 0 521 07862 8

First published 1970
Reprinted 1972 1974 1975

Printed in Great Britain
at the University Printing House, Cambridge
(Euan Phillips, University Printer)

Preface

This is the fifth of eight books primarily designed to cover a course suitable for those pupils who wish to take a C.S.E. examination on one of the reformed mathematics syllabuses.

The material is based upon the first four books of the O-level series, S.M.P. Books 1–4. The differences between this Main School series and the O-level series have been explained at length in the Preface to *Book A* as have the differences between the content of these two S.M.P. courses and that of the more traditional text.

It is important that pupils learn to cooperate with one another, and the Prelude to this book provides yet another excellent opportunity for pupils to work in small groups. Like the Preludes in the earlier books of the series, the one in *Book E* is essentially practical. Most of the work in the Prelude is on space filling, but there is a short section on three-dimensional drawing. The whole Prelude, while interesting in itself, provides an admirable background for the chapter on Volume later in the book.

In Chapter 1, no formal attempt is made to prove Pythagoras's theorem and indeed the word 'rule' is preferred to 'theorem', but several practical demonstrations of Pythagoras's rule are included. The problem of knowing the area of the square drawn on the hypotenuse, but of not knowing the length of the hypotenuse serves as a lead-in to the chapter, 'Square Roots'. Like the chapter on Pythagoras, this chapter has a practical bias and numerical work is fairly simple.

In this book, pupils graduate from the very elementary slide rule they made and used in *Book C* to one made from logarithmic graph paper. The importance of being able to estimate the size of an answer before actually using a slide rule is emphasized, and towards the end of the chapter, pupils learn how to find the square root of a number the easy way!

Teachers using this series of books will remember that so far in the course there has been no chapter on 'Sets'. A certain amount of set notation was introduced as it was needed in *Book A* and the notation has been used periodically ever since. In *Book E*, the use of punched cards provides motivation for the work that follows on the union and intersection of sets.

In *Book C*, pupils used matrices to describe networks, and in this book they learn how to add, subtract and multiply matrices. Then, in *Book F*, they will learn how matrices can be used in connection with work on relations and transformations and with further work on networks. Networks also come up again in this book, when Schlegel diagrams are considered.

Preface

Book E contains two chapters on a new topic—probability. Pupils find this a particularly interesting subject, and it is one which is likely to be of considerable everyday use in their future lives. The first chapter is concerned with the experimental approach to probability, while the second introduces the theoretical approach. There will be further work on probability later in the course.

The first of the chapters on solving equations algebraically and on the circle are contained in this book. The chapter on equations in *Book E* deals only with equations of the type $ax+b = c$ and $a(x+b) = c$ and shows how to solve them—initially by using flow diagrams. In the chapter on the circle, practical work is used to find approximate relations between the circumference and diameter and between the area and radius. π will be introduced in *Book G*. *Book E* also contains the introductory chapter of the course on trigonometry, which is based on the transformation of enlargement in a coordinate setting.

Answers to exercises are not printed at the end of this book but are contained in the companion Teacher's Guide which gives a detailed commentary on the pupil's text. In this series, the answers and commentary are interleaved with the text.

vi

Contents

Preface *page* v

Prelude 1
Filling space, 1; filling space with polyhedra, 4; drawing shapes, 9; deceptive drawings, 12

1 Right-angled triangles 13
The 3, 4, 5 triangle, 13; other triangles, 14; Pythagoras, 15; models to make, 21

2 Sets 25
Punched cards, 25; using diagrams, 28; sets of points and numbers, 31

3 Matrices 34
Storing information, 34; matrix addition, 36; combining row and column matrices 39; matrix multiplication, 42

4 Experiments 48
Experiments, 48; looking at results, 50

Interlude 53
Things are not what they seem, 53

Revision exercises 57
More practice with matrices, 61

5 Square roots 65
Square roots by calculation, 65; square roots by drawing, 70

6 Solving equations 74
Think of a number, 74; solving equations, 75

7 Probability 79
What is probability?, 79; equally likely outcomes, 81; theory and experiment, 88

8 The slide rule 93
Making and using a slide rule, 93; rough checks, 96; extending the slide rule, 97; accuracy, 99; use of a double scale, 102; the slide rule and fractions, 104

Contents

Interlude *page* 107
    Paper sizes 107

Revision exercises 110

9 Volume 114
    A reminder about area, 115; *units of volume,* 115; *volumes of cuboids,* 116; *volumes of prisms,* 119; *investigations,* 124

10 Enlargement 126
    Scale drawing, 126; *enlargement* 128; *using tables,* 130

11 The circle 136
    What do we mean by a circle?, 136; *technical terms,* 139; *measuring the circumference,* 140; *the area of a circle,* 144

12 Networks and polyhedra 150

Puzzle corner 156

Revision exercises 159

Prelude

1. FILLING SPACE

Suppose that you work for a company which has just produced a new washing powder called Sludge.

'Sludge, Sludge, fabulous Sludge, nothing quite like it for making dirt budge.'

Your job is to design a packet to contain this fabulous powder.

You could put it in a rectangular box or a circular tin:

Prelude

But perhaps a product such as Sludge deserves a more interesting shape. How about a dodecahedron?

Go to a large shop such as a supermarket and have a stroll round looking at the shapes of packets.

You will see many rectangular boxes (*cuboids*) and many circular tins (*cylinders*). Try to find some unusual packets. Look for shapes like these:

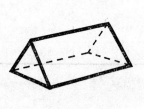 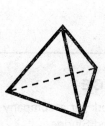 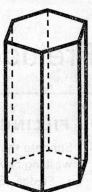

Did you find any dodecahedra? Why are manufacturers not keen to use them?

One reason why dodecahedra are not used is that they do not fit together very well. (There are other reasons too.)

Cuboids fit together neatly:

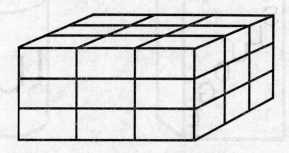

Filling space

Cylinders leave gaps but will fit tightly in a box:

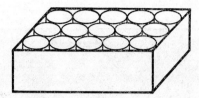

In previous work, you have studied the way flat shapes fit together (*tessellations*); for example:

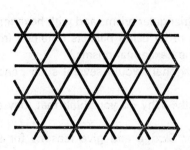
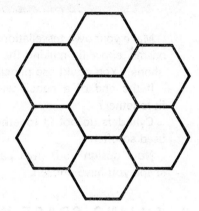

From these tessellations you can see that solid shapes in the form of triangular and hexagonal prisms will fit together:

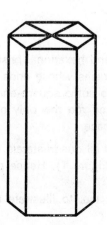
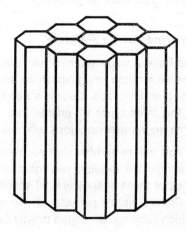

Prelude

The photograph shows part of the Giant's Causeway in Ireland. You can see that the rock has formed itself into prisms, many of which are hexagonal.

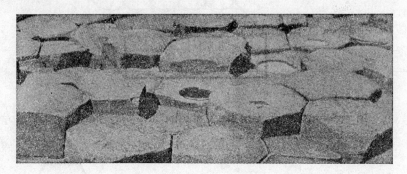

Make your own tessellations with triangular prisms and with hexagonal prisms, either by making the prisms or by using packets obtained from shops. (You could use pencils for the hexagonal prisms.)

If the end of a pencil was pentagonal, would several such pencils fit together?

Cylinders do not fit together exactly. Why do you think that they are used so often?

Now design your own packet for Sludge and give reasons for the shape you have chosen.

2. FILLING SPACE WITH POLYHEDRA

There are not many shapes which will fill space without leaving gaps. It is interesting to make models of some of the possibilities.

You will find that in most of the investigations which follow, it will help if you work in groups.

Investigation 1 : Prisms

We have already worked with triangular and hexagonal prisms. It does not take much imagination to see that prisms whose ends are square will also fill space. Design a net and make some square-ended prisms.

The three types of prisms just mentioned are the only ones whose ends are *regular* polygons which will fill space.

You will remember that all triangles and all quadrilaterals, whether or not they are regular, will tessellate (see Figure 1). Hence prisms with ends of these shapes will fill space.

Collect empty packets of, for example, tea, to illustrate the way in which rectangular-ended prisms fill space.

Filling space with polyhedra

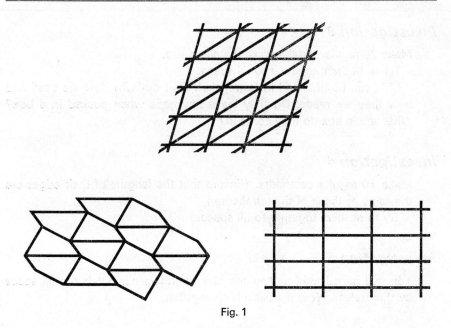

Fig. 1

Figure 2 shows a net for a quadrilateral-ended prism. You can design a net for a triangular-ended prism yourself.

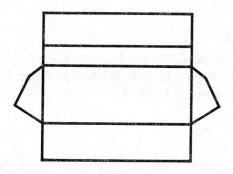

Fig. 2

Investigation 2: Cubes

A cube is a special case of a square-ended prism, and it is clear that cubes will fit together to fill space. It is helpful to have models of these for later use.

Make 20 identical cubes either by plaiting or by cutting out the nets.
Stick them together to fill space. Keep them somewhere safe as you will need them again in Investigation 7.

Prelude

Investigation 3

Make 20 regular tetrahedra of the same size.
Try to fit them together to fill space.
Milk can be obtained in 'tetrapaks'. Find out why they are used and how they are made. Do they leave any gaps when packed in a box? What shape box do they come in?

Investigation 4

Make 10 regular octahedra. (Ensure that the lengths of their edges are the same as those of the tetrahedra.)
Try to fit them together to fill space.

Investigation 5

You will have found that neither the tetrahedra nor octahedra fill space by themselves. Now try using both together.

Investigation 6

Investigate to see if the other regular polyhedra, that is, dodecahedra and icosahedra, fill space either by themselves or combined with other polyhedra.

Investigation 7

An interesting space-filler can be made in the following way.
Figure 3 shows a cube split up into 6 pyramids. The net for one of the pyramids is shown in Figure 4.

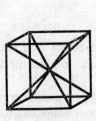
Fig. 3

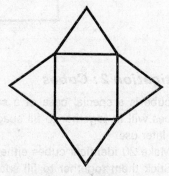
Fig. 4

Filling space with polyhedra

Make up six of these and sellotape their bases together as shown in Figure 5. You should find that the 6 pyramids fold up to make a cube.

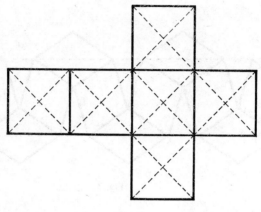

Fig. 5

Now instead of folding the 6 pyramids inwards, fold them outwards around another cube. You will get the shape shown in Figure 6. It is called a *rhombic dodecahedron* because its faces are rhombuses and there are twelve of them.

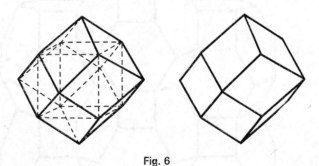

Fig. 6

Rhombic dodecahedra fill space. To see this, take the cubes stuck together in Investigation 2 and imagine that every alternate cube in the model is split into 6 pyramids. Now if each pyramid is stuck onto the cube next to it, the result will be a set of rhombic dodecahedra which all fit together.

Prelude

If you want to make a set of rhombic dodecahedra to satisfy yourself that they do fill space, it is easier to make them from their net (see Figure 7). The sizes of all angles are either $70\frac{1}{2}°$ or $109\frac{1}{2}°$ (approximately), and it is best to make the edges about 5 cm long.

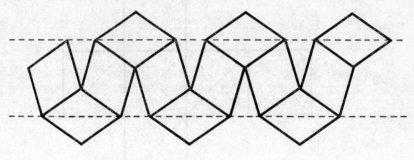

Fig. 7

Investigation 8

Another space-filler is the *truncated octahedron*. This is a regular octahedron with its corners cut off to form a polyhedron with 8 hexagonal faces and 6 square faces. Six of them are shown in Figure 8. The net is shown in Figure 9.

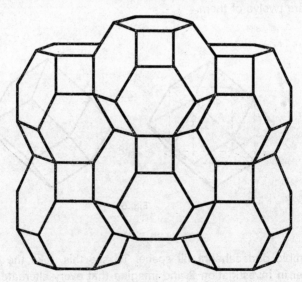

Fig. 8

Filling space with polyhedra

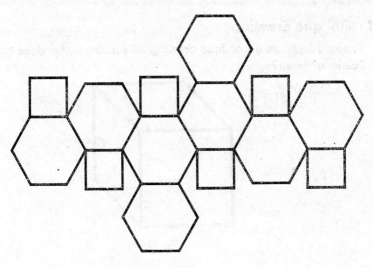

Fig. 9

3. DRAWING SHAPES

You have made some three-dimensional (solid) shapes. Now we are going to try to draw pictures of some of them in two dimensions, that is on a flat surface.

(*a*) Draw a picture of a cube. Pass it to your neighbour and see if he recognizes it. If not, try again.

Obtain a cube, preferably one made from straws and pipe-cleaners. Put it down in front of you so that it looks like your drawing. Are you satisfied with your drawing?

(*b*) Draw a picture of a chalk box.

Many people find that it is not easy to draw good pictures of three-dimensional objects. How did you get on? We shall now look at two methods often used in technical drawing.

Prelude

3.1 Oblique drawings

Figure 10 shows an oblique drawing of a cube. What does the word 'oblique' mean?

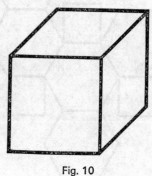

Fig. 10

The edges of a cube are all the same length. Are the edges of the drawing all the same length?

The faces of a cube are all squares. Are the faces of the drawing all squares?

The angles of a cube are all right-angles. Are the angles of the drawing all right-angles?

Figure 11 shows how to make an oblique drawing of a cube. Start with two squares, and then join up corresponding corners.

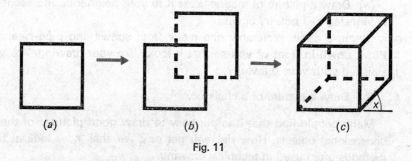

Fig. 11

In Figure 11 (c), the hidden edges have been drawn with dotted lines, but in Figure 10, the hidden edges are not shown at all. Which figure do you think gives the best representation of a cube?

Make some oblique drawings of a cube with the following variations:

(i) Try different lengths for the sloping edges to see which gives the best effect.

(ii) Try different sizes for the angle x between the horizontal and the sloping edges to see which gives the best effect.

Drawing shapes

Now draw pictures of (i) a brick, (ii) a four-legged rectangular table.

Do you think that your latest drawings are better than those you did at the beginning of Section 3?

3.2 Isometric drawings

You have used isometric paper before in constructing polyhedra and in making tessellations. Can you see a cube in Figure 12?
Place a model cube so that it looks like the diagram.

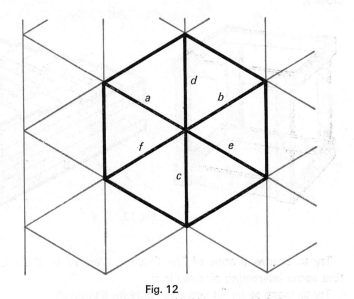

Fig. 12

Copy Figure 12 onto isometric paper drawing the lines lightly in pencil.

Draw thickly over the edges *a*, *b*, *c* so that they stand out and appear to be nearer. Draw the edges *d*, *e*, *f* with dotted lines so that they appear hidden. Which corner appears nearest to you, and which one farthest away?

Reverse the drawing system so that *a*, *b*, *c* are dotted (hidden) and *d*, *e*, *f* are dark. Does this give a different impression?

Isometric means equal length. Notice that each edge of the cube is represented by the same length of line.

The faces of a cube are all squares. What shape are the faces of the isometric drawing?

Prelude

The angles of a cube are all right-angles. What are the sizes of the angles of the isometric drawing?

Use isometric paper to make a drawing of a box which has a square end and whose length is twice its height.

4. DECEPTIVE DRAWINGS

You have done several drawings of objects, some of which have, no doubt, looked correct to you and others incorrect. Take a brief glance at Figure 13. Is there anything wrong with the drawings? Look more closely.

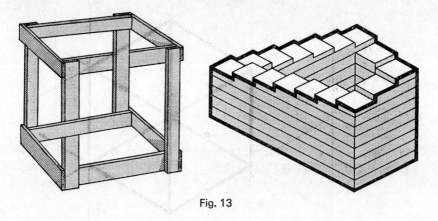

Fig. 13

Try to borrow a copy of *The Graphic Work of M. C. Escher*. You will find some interesting pictures in it.

Try to make some 'impossible' drawings yourself.

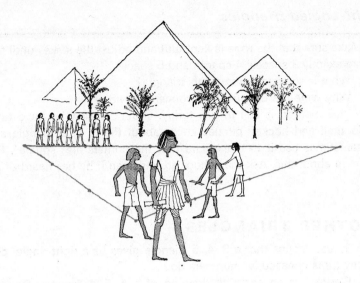

1. Right-angled triangles

1. THE 3, 4, 5 TRIANGLE

Thousands of years ago, surveyors in ancient Egypt could be seen setting out ropes as shown above. Indeed, they were known as 'rope stretchers'. If you look closely, you will see that they have made a triangle with three units on one side, four on the next and five on the third side.

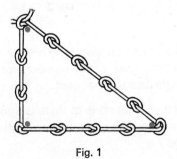

Fig. 1

Project

Obtain a long piece of thick string or thin rope and tie knots in it at equal intervals until you have thirteen knots. Get three sticks and go into the school field to stake out a triangle like the Egyptian one (see also Figure 1).

13

Right-angled triangles

Make sure that the rope is kept taut and adjust the stakes until the sides are exactly 3 spaces, 4 spaces and 5 spaces.

What is so special about this triangle?

Why were the Egyptians so concerned with it?

This method is still in use today. Groundsmen who have to set out football and hockey pitches know about the 3, 4, 5 triangle and they sometimes use it to make sure that the corners are 'square'. Find out more about this. Ask the school groundsman if he has heard of it.

2. OTHER TRIANGLES

Is it just chance that a 3, 4, 5 triangle gives us a right-angle, or would any three consecutive numbers do?

Figure 2 is an accurate drawing of a 4, 5, 6 triangle. Does it have a right-angle?

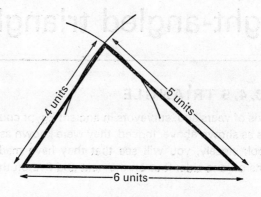

Fig. 2

Use a ruler and a pair of compasses to help you make accurate drawings of other triangles with consecutive numbers for lengths of sides and see if any of the triangles are right-angled.

You should discover that none of them is right-angled, and so it is not the fact that the numbers are consecutive which makes the 3, 4, 5 triangle such a special one.

What is it then?

3. PYTHAGORAS

It is hard to say who really did discover the answer to this problem. The Egyptians knew about it over 4000 years ago, and a Chinese document dated more than 1000 years B.C. refers to it. However, the important work seems to have been done by a Greek named Pythagoras about 530 B.C., and some years ago the Greek postal authorities issued a stamp to commemorate this man (see Figure 3).

Fig. 3

If you study this stamp it will show you what is so special about the 3, 4, 5 triangle.

What is the area of the large square?

What is the combined area of the other two squares?

See what would happen if you tried to do this with some of the other triangles you have considered. The 4, 5, 6 triangle would give:

6^2, that is, 36, for the large square, and

$4^2 + 5^2$, that is $16 + 25 = 41$, for the other two squares.

Try this with the other triangles you have drawn.

> Pythagoras stated that the area of the square drawn on the longest side of *any* right-angled triangle is equal to the *combined area* of the squares on the other two sides.

Right-angled triangles

Exercise A

1 Find the area of each of the squares in Figure 4. Add the two smallest together; does this equal the area of the largest?

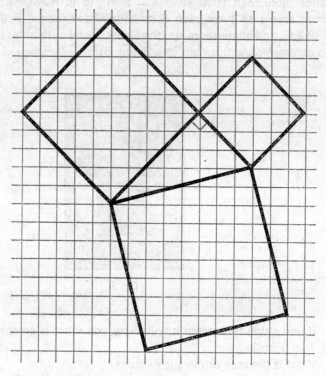

Fig. 4

2 Copy the triangle in Figure 5 onto graph paper and draw its associated squares. Find their areas and check Pythagoras's rule.

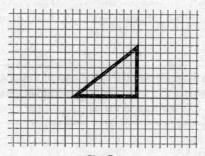

Fig. 5

3 Use graph paper to help you draw any right-angled triangle you like which is different from the one in Figure 5. Draw the squares on its sides, find their areas and check Pythagoras's rule.

4 Figure 6 shows a popular pattern for a tiled floor. Each tile is a right-angled isosceles triangle.

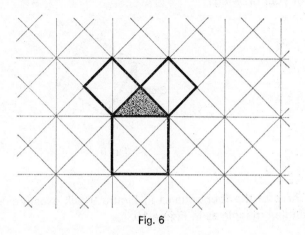

Fig. 6

Can you see an example of Pythagoras's rule here? How many tiles make the large square? How many tiles make the other two squares?

5 Check Pythagoras's rule with each of the triangles shown in the tiling pattern in Figure 7.

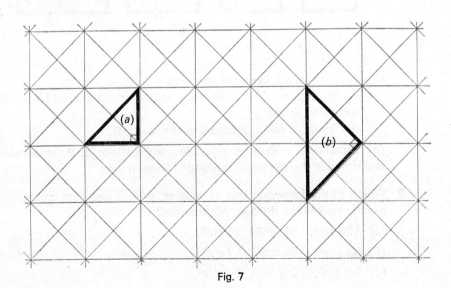

Fig. 7

Right-angled triangles

6 Draw Figure 8 *accurately* on plain paper. The square on the longest side should have area

$$2^2 + 3^2 = 4 + 9 = 13 \text{ cm}^2.$$

Measure the longest side, multiply it by itself and see how close to 13 your answer is. How does this provide a check on the accuracy of your drawing?

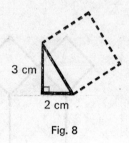

Fig. 8

7 Three of the four squares in Figure 9 will exactly surround a right-angled triangle as in Figure 10.

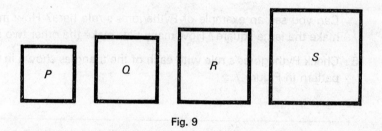

Fig. 9

(a) Find out which the three are:
   (i) by tracing them onto card, cutting them out and actually arranging them;
   (ii) by calculation (measure in centimetres).

(b) Experiment with other possible combinations of these squares and describe the triangles surrounded in each case.

8 Which of the following sets of squares, whose areas are given, will exactly surround a right-angled triangle?
   (a) 103 cm², 92 cm², 11 cm²;
   (b) 53 cm², 31 cm², 17 cm²;
   (c) 4·3 cm², 2·9 cm², 6·4 cm².

Pythagoras

9 Take A, B and C in Figure 10 as the areas of the squares for each part of this question, and find the missing number in each case.

(a) $A = 16$ cm², $B = 7$ cm², $C = ?$ cm²;

(b) $A = 28$ cm², $B = 17$ cm², $C = ?$ cm²;

(c) $A = 30$ cm², $B = ?$ cm², $C = 50$ cm²;

(d) $A = ?$ cm², $B = 167$ cm², $C = 225$ cm².

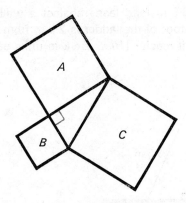

Fig. 10

10 In a triangle ABC, $AB = 6$ cm, $BC = 10$ cm and angle $ABC = 90°$. Find the length of AC by an accurate drawing and check your accuracy by calculation.

11 In a triangle LMN, $LM = 4$ cm, $LN = 6$ cm and angle $LMN = 90°$. Find the length of MN by drawing and check your accuracy by calculation.

12 Make an accurate drawing of a square of side 6 cm. Measure a diagonal and check its length by calculation.

13 Find the length of the side of a rhombus whose diagonals measure 8 cm and 6 cm, by making just a rough sketch and thinking.

Right-angled triangles

14 B is 12 km N.E. of A and C is 5 km N.W. of B. A straight road runs between A and B, and another between B and C (see Figure 11). How many kilometres would be saved by building a direct road from A to C?

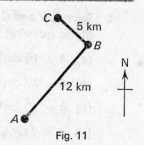

Fig. 11

15 A ladder 6·5 m long leans against a wall as shown in Figure 12. When the foot of the ladder is 2·5 m from the wall, how high up the wall does it reach? (*Hint:* work in units of $\frac{1}{2}$ m.)

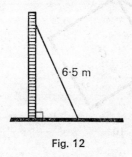

Fig. 12

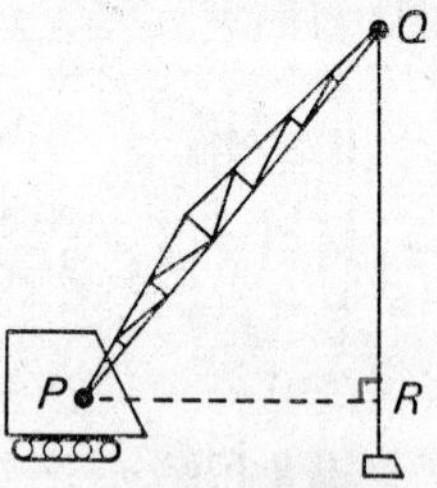

Fig. 13

16 Figure 13 shows the jib, PQ, of a crane of length 20 m. Find the height, QR, when the reach, PR, is: (*a*) 12 m; (*b*) 16 m.

17 In each entry of the following table, the lengths of the three sides of a triangle are given. Use a pair of compasses to help you make an accurate full size drawing of each one (see Figure 14).

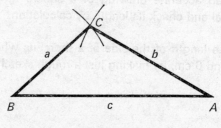

Fig. 14

Pythagoras

Copy and complete this table:

Sides (cm)			a^2	b^2	c^2	a^2+b^2	Is $a^2+b^2 = c^2$, $> c^2$, $< c^2$? (Put =, > or <)	Is angle BCA acute, right or obtuse?
a	b	c						
(i) 4	5	7						
(ii) 6	8	10						
(iii) 6	9	12						
(iv) 5	6	7						
(v) 5	12	13						
(vi) 4	8	10						
(vii) 5	9	6						

Can you make any comment about the results in the last two columns?

4. MODELS TO MAKE

1.

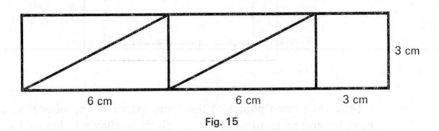

Fig. 15

Take two strips of card, 15 cm by 3 cm, mark them out as shown in Figure 15 and cut them up.

Draw accurately in the centre of a page the right-angled triangle shown in Figure 16.

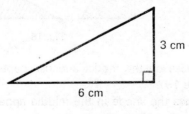

Fig. 16

21

Right-angled triangles 1

Take one set of five pieces, place the square against the 3 cm side of the triangle, and arrange the other four pieces to make a square against the 6 cm side.

Take the other set and assemble all *five* pieces to make a square on the longest side of the triangle.

What does this show you about the relation between the areas of the three squares surrounding the triangle?

Would this be true whatever size of strip you started with?

2. Take a sheet of cardboard and on it draw a square. Mark off an equal distance along each side and draw lines as in Figure 17, making two triangles.

What can you say about these two triangles?

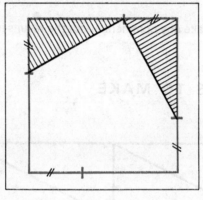

Fig. 17

Make two more copies of the same triangle from other pieces of card, leaving an extra piece for the pivots as shown in Figure 18.

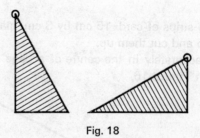

Fig. 18

Place these on the model and use paper fasteners to pivot them (see Figure 19).

What does the shape in the middle appear to be? Can you *prove* what you suspect?

Models to make

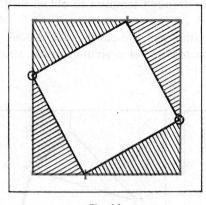
Fig. 19

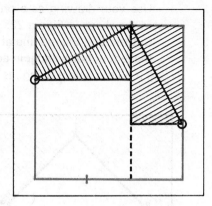
Fig. 20

Rotate each of the bottom triangles a quarter-turn, one clockwise, the other anticlockwise (see Figure 20).

In the first case the red square was composed of four equal right-angled triangles and another square in the middle. In the second case the red square is composed of the same four right-angled triangles, but instead of the square in the middle you have two squares at the bottom.

Explain how this demonstrates Pythagoras's rule.

3. Use squared paper to help you make an accurate copy of Figure 21, but to a larger scale.

Fig. 21

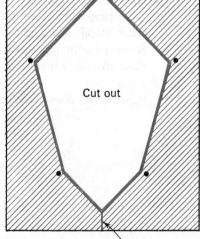

Fig. 22

Right-angled triangles

Use your drawing as a prick-through template to obtain on thick card the shape in Figure 22.

Cut through the red line at either end and cut the middle out.

Make four holes and join across with hat or shirring elastic, stretching it a little before fixing, as in Figure 23.

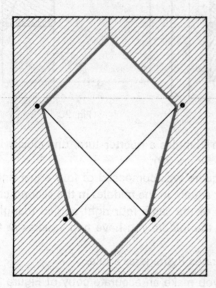
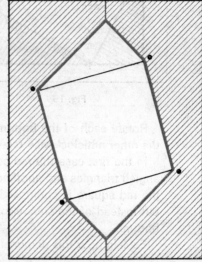

Fig. 23   Fig. 24

If you turn one side over you get the shape in Figure 24. Does the area of the hole change?

What shapes make up the hole in the first case?

What shapes make up the hole in the second case?

What sort of triangles are they? Are they all the same shape and size?

How are the two squares in Figure 23 related to the single square in Figure 24?

How are the sides of the squares related to the sides of the triangles?

2. Sets

1. PUNCHED CARDS

(*a*) You will know that in industry and commerce, computers are used for storing information, and for calculating such things as production figures and wages. The information is often fed into the computer on punched cards such as the one above. Try to get some of these cards and find out what the holes represent.

(*b*) Punched cards are also useful even if you do not have a computer. Figure 1 (over page) shows a card used at a school in Bristol for recording information about pupils in the school. In a large school, with many courses available, it is necessary to have information easily accessible, and punched cards such as these are helpful. There is one card for every pupil.

Sets 2

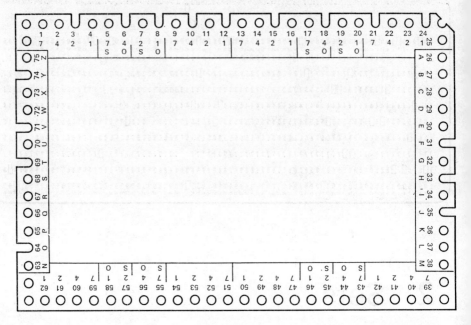

Fig. 1

Information is recorded on the card either by leaving a hole as it is, or by cutting a slit. For example, position 24 tells us whether the card refers to a boy or a girl. If a boy, the hole is left; if a girl, a slit is made. Position 15 is for Domestic Science. A hole indicates that the pupil takes this subject, a slit that the pupil does not take this subject.

The system works like this. Suppose the Headmaster wishes to know how many people in the fifth year do Woodwork. He takes the cards for the fifth year and pushes a long spike like a knitting needle through at the Woodwork position. Then he lifts it up and all the cards of pupils who do Woodwork are suspended on the spike.

If he then wants to know how many of these do Metalwork, he takes the cards he has just lifted off, removes the spike and pushes it in at the Metalwork position. By lifting it up, he then picks out those who do both of these subjects.

(c) It will be useful if you make up some punched cards yourself. If you cannot get cards like the one shown in Figure 1, it is possible to make your own. Your teacher will be able to tell you more about this.

You will have to decide what information you are going to store. You can only record information which can be represented by a hole or a slit.

Punched cards

For example, you could not record the answer to a question like 'How old are you?'. You would have to ask a question like 'Are you 14?'. It might be a good idea to record information which would be useful to your class teacher. Here, for example, is a card showing what subjects a certain pupil studies:

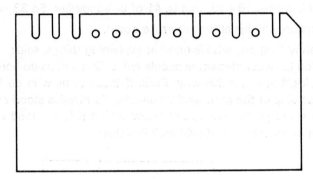

Fig. 2

You will need a code card to show what each position indicates:

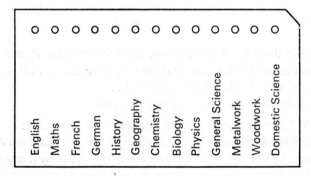

Fig. 3

What information does the card in Figure 2 tell you?

Explain how you would use a set of cards like this to sort out those people who do both Domestic Science and French.

How would you find those who do *either* Domestic Science *or* French? Your teacher will have some more questions for you which use the cards.

Sets

2. USING DIAGRAMS

(a) In Basil Brayne's class there are 33 pupils, and everyone does either French or Physics or both. Basil, who is not very bright, reckoned that he had proved that 33 = 41. He found that there were 18 doing French and 23 doing Physics. So he said:

'This makes 18 + 23, that is, 41 of us altogether. So 33 = 41.'

Explain why his argument broke down.

Penny Dropper, who is good at explaining things, said:

'You have counted some people twice. Some of us do both French and Physics. Look at it this way, Basil. If those of us who do French stood at this side of the room and those who do Physics stood over that side, then some people would not know which side to stand on. If we put them in the middle it would look like this:

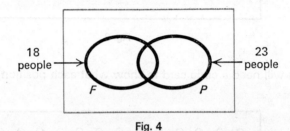

Fig. 4

Those in the middle are what we call F intersection P, or $F \cap P$.'

Basil asked her how many there would be in $F \cap P$. Penny did a quick calculation in her head and said 'Eight'.

Explain how she arrived at that number.

(b) To make sure that Basil understood, Penny said:

'Here is another problem. Suppose in another class there were 24 doing History, 15 doing Chemistry, and 9 doing both. How many people would there be in the class altogether, if everyone did one or other or both of these subjects?'

Basil drew a diagram like the one in Figure 5.

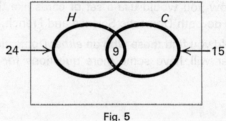

Fig. 5

Using diagrams

'What you want to know', he said, 'is how many there are in the union of H and C'.

He thought for a moment and then said 'There are 30 in $H \cup C$'. Was he right?

(c) Diagrams such as those Basil and Penny have drawn are useful in problems concerned with sets of things. They illustrate some of the results you will have found from your punched cards.

The shaded part of Figure 6 shows A intersection B, $(A \cap B)$,

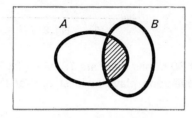

Fig. 6

and the shaded part of Figure 7 shows A union B, $(A \cup B)$.

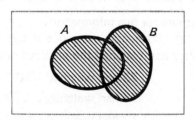

Fig. 7

Exercise A

1 In Basil's class of 33 pupils, everyone does either History or Chemistry or both. There are 24 doing History and 17 doing Chemistry. How many do both?

Sets 2

2. Figure 8 shows that in one class, 16 pupils do German, 18 do Woodwork and 5 do both. How many do German only? How many do Woodwork only? How many are there in the class if everyone does either German or Woodwork or both?

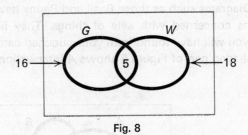

Fig. 8

3. In a class of 30 pupils there are 19 who play tennis and 16 who play hockey. If everyone plays at least one of these games how many play both?

4. Out of 17 girls in a class, 5 always walk to school, 7 use bicycles, and 9 use the bus service.

   $5+7+9$ is more than 17.

   What is the explanation?
   How many sometimes cycle and sometimes bus?

5. Draw a diagram for this information:
   In a survey 100 people were questioned about the BBC TV programmes they had watched the previous day.

   73 had watched BBC 1.
   28 had watched BBC 2.
   12 had watched both.

   How many people had not watched either?

6. All the inhabitants of a certain town on the French–German border speak either French or German or both. If 64% speak French and 58% speak German, what percentage speak both languages?

7. The number of people in $A \cup B$ is 69, the number in A is 41, and the number in B is 53. Draw a diagram with two overlapping curves representing sets A and B.
   How many people are there in $A \cap B$?

8. The number of people in C is 12, in D is 11, and in $C \cap D$ is 4. Draw a diagram with two overlapping curves representing sets C and D.
   How many are there in $C \cup D$?

30

Using diagrams

9 Invent some questions like the ones in this exercise which refer to your own class.

10 Make six rough copies of Figure 9 and on separate diagrams, shade the following sets:
(a) A ∩ B; (b) A ∪ B; (c) B ∩ C;
(d) B ∪ C; (e) C ∩ A; (f) C ∪ A.

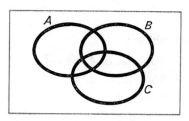

Fig. 9

3. SETS OF POINTS AND NUMBERS

We have been finding the intersection and union of sets of *people*. The same ideas apply to other sets, such as sets of points and sets of numbers. Here are some examples.

(a) On squared paper, mark the set of points, A, such that $x = 3$.
Mark also the set of points, B, such that $x = y$.
Indicate A ∩ B.

This is why the word 'intersection' is used: the lines *intersect* or cross each other at the point A ∩ B.

(b) On squared paper, shade the set of points, C, such that $x > 3$.
In another colour, shade the set of points, D, such that $y > 4$.
Which set on your diagram is C ∩ D? Which set is C ∪ D?

(c) Suppose E is the set of multiples of two, up to twenty, that is, E is the set   2, 4, 6, 8, 10, 12, 14, 16, 18, 20.

We usually write this as:

$$E = \{2, 4, 6, 8, 10, 12, 14, 16, 18, 20\}.$$

The curly brackets are short for 'is the set of'.
Give another description of F if

$$F = \{3, 6, 9, 12, 15, 18\}.$$

Write down the members of the set E ∩ F.
Write down the members of the set E ∪ F.

Sets

(d) Give another description of
  (i) {a, e, i, o, u};
  (ii) {b, c, d, f, g, h, j, k, l, m, n, p, q, r, s, t, v, w, x, y, z}.

For short, call (i) V and (ii) C.
What is $V \cup C$?
How many members in $V \cap C$?

Exercise B

1 Copy and complete:
  (a) {1, 3, 7, 10} ∩ {2, 3, 4, 6} = { };
  (b) {1, 3, 7, 10} ∪ {2, 3, 4, 6} = { };
  (c) {a, b, c} ∪ {d, e, f} = { };
  (d) {a, b, c} ∪ {c, d, e, f} = { };
  (e) {7, 9, } ∩ {5, , 2, 3} = {9, 3}.

2 On squared paper, mark the set of points, A, such that $x = 2$.
   Mark also the set, B, such that $y = 3$.
   What can you say about $A \cap B$?

3 On squared paper, shade the set of points, C, such that $x > 4$. In another colour, shade the set of points, D, such that $y < 2$.
   Which set is $C \cap D$?
   Which set is $C \cup D$?

4 E is the region $x < 3$, F the region $y > 1$, and G the region $x > 5$. Using different colours shade them in.
   Show on separate diagrams, the regions
     (a) $E \cap F$;
     (b) $F \cap G$;
     (c) $G \cap E$;
     (d) $E \cup F$.

5 List the members of the following sets:
     H = {multiples of 2 which are less than 40},
     I = {multiples of 3 which are less than 40},
     J = {multiples of 5 which are less than 40}.

   Find
     (a) $H \cap I$;
     (b) $I \cap J$;
     (c) $J \cap H$.

Sets of points and numbers

6 List the members of the following sets:

$$K = \{\text{factors of 15}\},$$
$$L = \{\text{factors of 18}\},$$
$$M = \{\text{factors of 12}\},$$
$$N = \{\text{prime numbers less than 20}\}.$$

Find:
(a) $K \cap N$;
(b) $L \cap N$;
(c) $M \cap N$.

7 List the members of the following sets:

$$P = \{\text{prime numbers less than 50}\},$$
$$R = \{\text{rectangle numbers less than 50}\},$$
$$S = \{\text{square numbers less than 50}\},$$
$$T = \{\text{triangle numbers less than 50}\}.$$

Find:
(a) $S \cap R$;
(b) $S \cap T$;
(c) $P \cap T$;
(d) $R \cap T$.

8 Make two copies of Figure 10 and on one shade the set

$$A \cap (B \cup C)$$

and on the other

$$(A \cap B) \cup (A \cap C).$$

What do you notice?

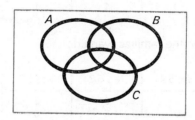

Fig. 10

9 Make two more copies of Figure 10 and on one shade the set

$$A \cup (B \cap C)$$

and on the other

$$(A \cup B) \cap (A \cup C).$$

What do you notice?

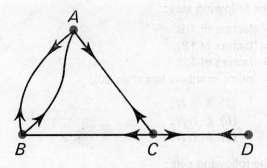

3. Matrices

1. STORING INFORMATION

(a) Mrs Brayne orders all the milk for the four families living in her block of flats. The table below refers to the number of bottles of milk delivered one Monday.

	No. 5A	No. 5B	No. 5C	No. 5D	
Gold	2	0	2	1	
Red	0	2	1	3	MONDAY
Silver	0	1	1	1	

On Tuesday, the families wanted:

	No. 5A	No. 5B	No. 5C	No. 5D	
Gold	1	0	2	2	
Red	0	3	3	2	TUESDAY
Silver	0	1	0	2	

Mrs Brayne needs to know how much milk each family has. Instead of writing out charts in full like the ones above, she stores the information in number parcels. On Monday, she wrote:

$$\begin{pmatrix} 2 & 0 & 2 & 1 \\ 0 & 2 & 1 & 3 \\ 0 & 1 & 1 & 1 \end{pmatrix}.$$

Storing information

She knows what each number in this parcel means because, for example, the 2nd row always refers to red top milk and the 3rd column to No. 5 C.

What number parcel did she write for Tuesday?

Number parcels are called *matrices* and we have used them before for storing information. For example, the matrix at the head of this chapter describes a network. What information does it give? What does the number 2 tell you?

(*b*) We do not know what Mrs Brayne wrote for Wednesday. But we do know that the matrix has the same shape as the others. How many rows has it? How many columns?

A matrix which has 3 *rows* and 4 *columns* is called a 3 by 4 matrix. We always write the *row number first*. We call 3 by 4 the *order* of the matrix.

(*c*) Explain why the order of the matrix

$$\begin{pmatrix} 1 & 2 & 3 & 4 & 5 \\ 0 & 3 & 6 & 5 & 7 \end{pmatrix} \text{ is 2 by 5.}$$

(*d*) State the orders of the following matrices:

(i) $\begin{pmatrix} 6 & 4 \\ 1 & -2 \\ 3 & 5 \\ -1 & 1 \end{pmatrix}$;   (ii) $\begin{pmatrix} 1 & 3 & 0 \\ 5 & 1 & 6 \end{pmatrix}$.

(*e*) State the order of the matrix $\begin{pmatrix} 3 \\ 8 \\ 1 \end{pmatrix}$. We sometimes call this a *column matrix* because all the numbers are in one column.

(*f*) What is the order of the matrix (2 4 −1 5)? What name do you think we give to a matrix when all the numbers are in one row?

(*g*) What is the order of the matrix at the head of this chapter? We call it a *square matrix*. Why? Write down a square matrix which has only two columns.

Matrices

Exercise A

1. Write down:
   - (a) a 2 by 4 matrix;
   - (b) a 4 by 3 matrix;
   - (c) a 5 by 2 matrix;
   - (d) a 2 by 5 matrix.

2. State the orders of the following matrices:

   (a) $\begin{pmatrix} 2 \\ 5 \end{pmatrix}$; (b) $(4\ 6\ 1)$; (c) $\begin{pmatrix} 2 & 3 \\ 1 & 7 \end{pmatrix}$;

   (d) $\begin{pmatrix} 7 & 6 & 2 & 1 \\ 3 & 0 & -5 & 9 \end{pmatrix}$; (e) $\begin{pmatrix} 8 & 1 & 2 \\ -1 & 3 & 5 \\ 6 & 4 & 0 \end{pmatrix}$;

   (f) $(7\ 5\ 9\ 4\ 3)$; (g) $\begin{pmatrix} 2 \\ 7 \\ 6 \\ 3 \end{pmatrix}$; (h) $(4\ -3)$.

   Which of these matrices are (i) column matrices, (ii) row matrices, (iii) square matrices?

3. Three shops hold the following stocks of 'pop' records. Shop A has 60 L.P.s, 87 E.P.s and 112 singles; shop B has 103 L.P.s, 41 E.P.s and 58 singles; shop C sells only singles and has 147 of them. Write this information in matrix form and state the order of the matrix.

4. One Sunday in November the following information was obtained from a newspaper: at the top of the first division, Liverpool had played 18 games, won 11, drawn 4 and lost 3; Everton had played 18, won 10, drawn 6 and lost 2; Leeds had played 17, won 10, drawn 5 and lost 2. Write this information in a 3 by 4 matrix.

2. MATRIX ADDITION

(a) On Monday, Mrs Brayne ordered:

$$\begin{pmatrix} 2 & 0 & 2 & 1 \\ 0 & 2 & 1 & 3 \\ 0 & 1 & 1 & 1 \end{pmatrix}$$

and on Tuesday:

$$\begin{pmatrix} 1 & 0 & 2 & 2 \\ 0 & 3 & 3 & 2 \\ 0 & 1 & 0 & 2 \end{pmatrix}.$$

How much gold top milk did each family order on Monday and Tuesday together?

Matrix addition

Copy and complete the 3 by 4 matrix below to show how much of each type of milk each flat had on Monday and Tuesday together:

$$\begin{pmatrix} 3 & 0 & 4 & 3 \\ 0 & & & \\ 0 & & & \end{pmatrix}.$$

To complete this matrix you had to add the pairs of numbers which have corresponding positions in the matrices for Monday and Tuesday. For example, to find the bottom right-hand number, you had to add the bottom right-hand numbers 1 and 2 to obtain 3.

This way of combining two matrices to form a third is called *matrix addition* and we write:

$$\begin{pmatrix} 2 & 0 & 2 & 1 \\ 0 & 2 & 1 & 3 \\ 0 & 1 & 1 & 1 \end{pmatrix} + \begin{pmatrix} 1 & 0 & 2 & 2 \\ 0 & 3 & 3 & 2 \\ 0 & 1 & 0 & 2 \end{pmatrix} = \begin{pmatrix} 3 & 0 & 4 & 3 \\ 0 & 5 & 4 & 5 \\ 0 & 2 & 1 & 3 \end{pmatrix}.$$

We can *add* two matrices only if they have the *same order*, that is, if they are the same shape. Why?

(*b*) We often label matrices with capital letters in heavy type like this:

$$\mathbf{A} = \begin{pmatrix} 3 & 4 \\ 5 & 6 \\ 4 & 5 \end{pmatrix} \quad \text{and} \quad \mathbf{B} = \begin{pmatrix} 2 & 3 \\ 3 & 0 \\ 4 & 2 \end{pmatrix}.$$

Then we can write

$$\mathbf{A} + \mathbf{B} = \begin{pmatrix} 3+2 & 4+3 \\ 5+3 & 6+0 \\ 4+4 & 5+2 \end{pmatrix} = \begin{pmatrix} 5 & 7 \\ 8 & 6 \\ 8 & 7 \end{pmatrix}.$$

What is the order of (i) **A**; (ii) **B**; (iii) **A** + **B**?

(*c*) We can also subtract **B** from **A** if we wish and write

$$\mathbf{A} - \mathbf{B} = \begin{pmatrix} 3-2 & 4-3 \\ 5-3 & 6-0 \\ 4-4 & 5-2 \end{pmatrix} = \begin{pmatrix} 1 & 1 \\ 2 & 6 \\ 0 & 3 \end{pmatrix}.$$

If

$$\mathbf{C} = \begin{pmatrix} 9 & 0 \\ 3 & 4 \end{pmatrix} \quad \text{and} \quad \mathbf{D} = \begin{pmatrix} 4 & 0 \\ 1 & 2 \end{pmatrix},$$

find (i) **C** + **D**; (ii) **D** + **C**; (iii) **C** + **C**; (iv) **C** − **D**.

Matrices 3

Exercise B

1. Let $P = \begin{pmatrix} 3 & 4 & 2 \\ 4 & 5 & 6 \end{pmatrix}$ and $Q = \begin{pmatrix} 1 & 1 & 0 \\ 1 & 2 & 3 \end{pmatrix}$.

   Find (a) P+Q; (b) Q+P. What do your answers suggest?

2. Write down any two matrices **A** and **B** which have the *same order*. Find (a) A+B; (b) B+A. Do you think that matrix addition is commutative?

3. Let
   $T = \begin{pmatrix} 1 & 2 \\ 3 & 1 \\ 5 & 6 \end{pmatrix}$, $U = \begin{pmatrix} 1 & 0 \\ 2 & 4 \\ 3 & 2 \end{pmatrix}$ and $V = \begin{pmatrix} 5 & 2 \\ 3 & 2 \\ 4 & 1 \end{pmatrix}$.

   Find (a) (T+U)+V; (b) T+(U+V). What do your answers suggest? Is there any confusion if we write T+U+V?

4. Write down any three matrices **A**, **B** and **C** which have the *same order*. Find (a) (A+B)+C; (b) A+(B+C). Do you think that matrix addition is associative?

5. Let $E = \begin{pmatrix} 1 & 2 & 4 \\ 0 & 5 & 3 \end{pmatrix}$.

   (a) Find E+E.
   (b) We write E+E as 2E. How would you write

       (i) E+E+E; (ii) E+E+E+E+E?

   (c) Find: (i) 3E; (ii) 5E.

6. Let $G = \begin{pmatrix} 11 & 3 \\ 4 & 7 \end{pmatrix}$ and $H = \begin{pmatrix} 5 & 0 \\ 2 & 1 \end{pmatrix}$.

   Find: (a) G+H; (b) G−H; (c) 2G;
        (d) 4H; (e) G+4H; (f) 2G−4H.

7. In the first match of the cricket season Brown bowled 12 overs, of which 4 were maidens, and he took 5 wickets for 25 runs. In the second match he took 3 wickets in 17 overs, of which 3 were maidens, for 52 runs.

   Write the figures for each match as 1 by 4 matrices. (Be careful about the order in which you write the numbers!) Now write down the 1 by 4 matrix which shows Brown's figures for the two matches combined.

Matrix addition

8 Let

$$J = \begin{pmatrix} 3 & 2 & 2 \\ 4 & 5 & 9 \end{pmatrix}, \quad K = \begin{pmatrix} 4 & 5 \\ 2 & 3 \\ 6 & 7 \end{pmatrix}, \quad L = \begin{pmatrix} 7 & 8 & 9 \\ 10 & 11 & 12 \\ 13 & 14 & 15 \end{pmatrix},$$

$$M = \begin{pmatrix} 1 & 1 \\ 0 & 1 \\ 0 & 1 \end{pmatrix} \quad \text{and} \quad N = \begin{pmatrix} 5 & 4 & 7 \\ 8 & 0 & 9 \end{pmatrix}.$$

Find where possible:
(a) **K + M**; (b) **J + L**; (c) **J + M**; (d) **J + N**;
(e) **N + J**; (f) **2N**; (g) **N + K**; (h) **5N**.

3. COMBINING ROW AND COLUMN MATRICES

So far we have combined matrices by adding and subtracting them. We shall now see that sometimes it is sensible to combine them in a different way.

In an athletics match between three schools, points were given for 1st, 2nd, 3rd and 4th places. This table shows the results for school A:

	1sts	2nds	3rds	4ths
School A	5	2	4	4

and this shows the number of points awarded for each place:

	1sts	2nds	3rds	4ths
Number of points	5	3	2	1

Find the total number of points gained by school A.

We can write both sets of information as matrices:

$$(5 \ 2 \ 4 \ 4) \quad \text{and} \quad (5 \ 3 \ 2 \ 1).$$

Do you agree that

$(5 \ 2 \ 4 \ 4)$ can be combined with $(5 \ 3 \ 2 \ 1)$ to give 43?

This way of combining matrices is called *matrix multiplication* and we usually write it like this:

$$(5 \ 2 \ 4 \ 4) \begin{pmatrix} 5 \\ 3 \\ 2 \\ 1 \end{pmatrix} = 25 + 6 + 8 + 4 = 43.$$

The *first* matrix is written as a *row* matrix and the *second* matrix as a *column* matrix. It is not necessary to put brackets round 43 since it is a single element. Notice also that there is no multiplication sign between the two matrices which we wish to combine.

Matrices

School *B* gained 0 1sts, 8 2nds, 7 3rds and 7 4ths. We can find the total number of points gained by working out

$$(0\ 8\ 7\ 7) \begin{pmatrix} 5 \\ 3 \\ 2 \\ 1 \end{pmatrix} = 0 + 24 + 14 + 7 = 45.$$

We can multiply matrices in this way only if the number of *elements* or numbers in the row matrix is equal to the number of elements in the column matrix.

We cannot write

$$(8\ 7\ 7) \begin{pmatrix} 5 \\ 3 \\ 2 \\ 1 \end{pmatrix}$$

because it has no meaning. Why not?

School *C* gained 7 1sts, 2 2nds, 1 3rd, and 1 4th. Write this information as a row matrix. By multiplying two matrices together, find the total number of points gained by school *C*. Set your work out carefully. Which school won the match?

Exercise C

1 Chelsea have won 10 games, drawn 6 and lost 2. We can show this by the row matrix (10 6 2).

For a win they get 2 points, for a draw 1 point and for losing 0 points. We can show this by the column matrix $\begin{pmatrix} 2 \\ 1 \\ 0 \end{pmatrix}$.

Multiply the two matrices to find how many points Chelsea have.

2 Work out:

(a) $(3\ 2\ 1) \begin{pmatrix} 4 \\ 5 \\ 6 \end{pmatrix}$; (b) $(4\ 5\ 6\ 7) \begin{pmatrix} 1 \\ 2 \\ 3 \\ 4 \end{pmatrix}$;

(c) $(10\ 20) \begin{pmatrix} 20 \\ 30 \end{pmatrix}$; (d) $(1\ 0\ 0\ 0) \begin{pmatrix} 1 \\ 2 \\ 3 \\ 4 \end{pmatrix}$;

(e) $(5\ 2\ 7) \begin{pmatrix} 1 \\ 1 \\ 1 \end{pmatrix}$; (f) $(4\ 5) \begin{pmatrix} 2 \\ 3 \end{pmatrix}$.

Combining row and column matrices

3 In a traffic census the following information was obtained:

	1 in car	2 in car	3 in car	4 in car
Number of cars	96	40	20	9

Write this information as a row matrix.

Use the column matrix $\begin{pmatrix} 1 \\ 2 \\ 3 \\ 4 \end{pmatrix}$ to find the total number of people carried in all the cars.

4 The way in which the first XV scored in the first match of the season is as follows:

	Tries	Conversions	Penalty goals
Match 1	5	1	3

3 points are awarded for a try, 2 for a conversion and 3 for a penalty goal. Write out the points scheme as a column matrix and use it and another matrix to find how many points the team scored in their first match.

	Tries	Conversions	Penalty goals
Match 2	3	2	1
Match 3	2	0	3

Do the same for the second and third matches.

5 Work out the following multiplications *if they have a meaning:*

(a) $(5\ 7)\begin{pmatrix} 6 \\ 8 \end{pmatrix}$;

(b) $(2\ 3\ 4)\begin{pmatrix} 5 \\ 6 \\ 7 \\ 8 \end{pmatrix}$;

(c) $(0\ 0\ 0)\begin{pmatrix} 0 \\ 0 \\ 0 \end{pmatrix}$;

(d) $(4\ 5)\begin{pmatrix} 2 \\ 3 \\ 0 \end{pmatrix}$;

(e) $(0\ 0\ 0\ 0)\begin{pmatrix} 0 \\ 0 \end{pmatrix}$;

(f) $(3\ 4\ 5\ 6)\begin{pmatrix} 4 \\ 3 \\ 2 \\ 1 \end{pmatrix}$.

Matrices 3

4. MATRIX MULTIPLICATION

We can use a 3 by 4 matrix to show the results of the athletics match for all three schools:

$$\begin{array}{c} \text{School } A \\ \text{School } B \\ \text{School } C \end{array} \begin{pmatrix} \overset{\text{1sts}}{5} & \overset{\text{2nds}}{2} & \overset{\text{3rds}}{4} & \overset{\text{4ths}}{4} \\ 0 & 8 & 7 & 7 \\ 7 & 2 & 1 & 1 \end{pmatrix}.$$

We know that

$$(5 \ 2 \ 4 \ 4) \begin{pmatrix} 5 \\ 3 \\ 2 \\ 1 \end{pmatrix} = 43,$$

so school A gained 43 points.
Also,

$$(0 \ 8 \ 7 \ 7) \begin{pmatrix} 5 \\ 3 \\ 2 \\ 1 \end{pmatrix} = 45$$

and

$$(7 \ 2 \ 1 \ 1) \begin{pmatrix} 5 \\ 3 \\ 2 \\ 1 \end{pmatrix} = 44,$$

so school B gained 45 points and school C gained 44 points.

We can stack these three multiplications together and write:

$$\begin{array}{c} A \\ B \\ C \end{array} \begin{pmatrix} \overset{\text{1sts}}{5} & \overset{\text{2nds}}{2} & \overset{\text{3rds}}{4} & \overset{\text{4ths}}{4} \\ 0 & 8 & 7 & 7 \\ 7 & 2 & 1 & 1 \end{pmatrix} \begin{pmatrix} 5 \\ 3 \\ 2 \\ 1 \end{pmatrix}\begin{matrix} \text{1sts} \\ \text{2nds} \\ \text{3rds} \\ \text{4ths} \end{matrix} = \begin{array}{c} A \\ B \\ C \end{array}\begin{pmatrix} \overset{\text{points}}{43} \\ 45 \\ 44 \end{pmatrix}.$$

The red labels are not necessary. They have been put in to help you to see how the three separate multiplications have been stacked together to form a single multiplication.

Copy and complete the following multiplication to find the number of points each school would have gained if 8 points had been given for a 1st, 5 points for a 2nd, 3 points for a 3rd and 1 point for a 4th.

$$\begin{pmatrix} 5 & 2 & 4 & 4 \\ 0 & 8 & 7 & 7 \\ 7 & 2 & 1 & 1 \end{pmatrix} \begin{pmatrix} 8 \\ 5 \\ 3 \\ 1 \end{pmatrix} = \begin{pmatrix} 66 \\ \ \\ \ \end{pmatrix}.$$

Matrix multiplication

Which school would have won the match if this scoring system had been used?

We can now stack the two scoring systems side by side and write:

$$\begin{pmatrix} 5 & 2 & 4 & 4 \\ 0 & 8 & 7 & 7 \\ 7 & 2 & 1 & 1 \end{pmatrix} \begin{pmatrix} 5 & 8 \\ 3 & 5 \\ 2 & 3 \\ 1 & 1 \end{pmatrix} = \begin{pmatrix} 43 & 66 \\ 45 & 68 \\ 44 & 70 \end{pmatrix}.$$

When we multiply two matrices we combine each row of the first matrix with each column of the second matrix. Let us see this happen again for the matrices

$$\mathbf{A} = \begin{pmatrix} 3 & 2 & 4 \\ 4 & 2 & 1 \end{pmatrix} \text{ and } \mathbf{B} = \begin{pmatrix} 1 & 1 \\ 2 & 0 \\ 1 & 3 \end{pmatrix}.$$

The first row of **A** combines with the columns of **B** to give

$$(3 \ 2 \ 4) \begin{pmatrix} 1 \\ 2 \\ 1 \end{pmatrix} = 11 \text{ and } (3 \ 2 \ 4) \begin{pmatrix} 1 \\ 0 \\ 3 \end{pmatrix} = 15.$$

The second row of **A** combines with the columns of **B** to give

$$(4 \ 2 \ 1) \begin{pmatrix} 1 \\ 2 \\ 1 \end{pmatrix} = 9 \text{ and } (4 \ 2 \ 1) \begin{pmatrix} 1 \\ 0 \\ 3 \end{pmatrix} = 7.$$

Stacking the rows gives

$$\begin{pmatrix} 3 & 2 & 4 \\ 4 & 2 & 1 \end{pmatrix} \begin{pmatrix} 1 \\ 2 \\ 1 \end{pmatrix} = \begin{pmatrix} 11 \\ 9 \end{pmatrix} \text{ and } \begin{pmatrix} 3 & 2 & 4 \\ 4 & 2 & 1 \end{pmatrix} \begin{pmatrix} 1 \\ 0 \\ 3 \end{pmatrix} = \begin{pmatrix} 15 \\ 7 \end{pmatrix},$$

and now stacking the columns, we write

$$\mathbf{AB} = \begin{pmatrix} 3 & 2 & 4 \\ 4 & 2 & 1 \end{pmatrix} \begin{pmatrix} 1 & 1 \\ 2 & 0 \\ 1 & 3 \end{pmatrix} = \begin{pmatrix} 11 & 15 \\ 9 & 7 \end{pmatrix}.$$

What is the order of **A**? The 3 tells us the number of elements in each row of **A**. What is the order of **B**? This time the 3 tells us the number of elements in each column of **B**. Since these two numbers are equal, we can combine the rows of **A** with the columns of **B** and work out **AB**. If the numbers were not equal, **AB** would not have a meaning.

$$\mathbf{BA} = \begin{pmatrix} 1 & 1 \\ 2 & 0 \\ 1 & 3 \end{pmatrix} \begin{pmatrix} 3 & 2 & 4 \\ 4 & 2 & 1 \end{pmatrix}.$$

Does this have a meaning?

Matrices

If $C = \begin{pmatrix} 2 & 0 \\ 3 & 5 \end{pmatrix}$ and $D = \begin{pmatrix} 1 & 4 \\ 7 & 3 \\ 2 & 5 \end{pmatrix}$,

does **CD** have a meaning? Does **DC** have a meaning?

Copy and complete the following multiplications. The red loops in the first one are to remind you how to work out the answers. Underneath each matrix write its order.

(a) $\begin{pmatrix} 2 & 4 & 0 \\ 3 & 6 & 1 \end{pmatrix} \begin{pmatrix} 1 & 0 & 1 \\ 1 & 2 & 3 \\ 0 & 5 & 4 \end{pmatrix} = \begin{pmatrix} 6 & 8 & 14 \\ 9 & & \end{pmatrix}$;

(b) $\begin{pmatrix} 7 & 2 \\ 1 & 0 \\ 4 & 1 \end{pmatrix} \begin{pmatrix} 1 & 3 & 2 & 1 \\ 4 & 5 & 0 & 3 \end{pmatrix} = \begin{pmatrix} 15 & 31 & 14 & 13 \\ & & & 1 \\ & & & 7 \end{pmatrix}$;

(c) $\begin{pmatrix} 6 & 3 \\ 2 & 4 \end{pmatrix} \begin{pmatrix} 1 & 5 \\ 1 & 3 \end{pmatrix} = \begin{pmatrix} 9 & \\ & 22 \end{pmatrix}$;

(d) $\begin{pmatrix} 8 & 1 & 5 \\ 2 & 0 & 7 \end{pmatrix} \begin{pmatrix} 5 & 0 & 2 & 0 \\ 1 & 4 & 1 & 0 \\ 9 & 3 & 2 & 1 \end{pmatrix} = \begin{pmatrix} & & 27 & 5 \\ 73 & 21 & & \end{pmatrix}$;

(e) $\begin{pmatrix} 5 & 0 \\ 2 & 3 \\ 1 & 5 \\ 3 & 7 \end{pmatrix} \begin{pmatrix} 1 & 2 & 0 & 3 & 2 \\ 6 & 1 & 4 & 0 & 1 \end{pmatrix} = \begin{pmatrix} 5 & & & & \\ & 7 & 12 & & \\ & & 20 & 3 & \\ & & & & 13 \end{pmatrix}$.

Look again at the orders which you have written underneath the matrices and compare them with this domino pattern:

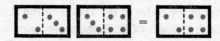

Matrix multiplication

Exercise D

1. Work out:

    (a) $(3 \ 4 \ 5)\begin{pmatrix} 1 \\ 2 \\ 3 \end{pmatrix}$;

    (b) $(1 \ 3 \ 6)\begin{pmatrix} 1 \\ 2 \\ 3 \end{pmatrix}$;

    (c) $\begin{pmatrix} 3 & 4 & 5 \\ 1 & 3 & 6 \end{pmatrix}\begin{pmatrix} 1 \\ 2 \\ 3 \end{pmatrix}$;

    (d) $\begin{pmatrix} 3 & 4 & 5 \\ 1 & 3 & 6 \end{pmatrix}\begin{pmatrix} 1 & 2 \\ 2 & 5 \\ 3 & 4 \end{pmatrix}$.

2. Work out:

    (a) $(7 \ 1 \ 5)\begin{pmatrix} 2 \\ 8 \\ 0 \end{pmatrix}$;

    (b) $(7 \ 1 \ 5)\begin{pmatrix} 3 \\ 6 \\ 2 \end{pmatrix}$;

    (c) $(7 \ 1 \ 5)\begin{pmatrix} 2 & 3 \\ 8 & 6 \\ 0 & 2 \end{pmatrix}$;

    (d) $\begin{pmatrix} 7 & 1 & 5 \\ 0 & 4 & 3 \\ 1 & 5 & 6 \end{pmatrix}\begin{pmatrix} 2 & 3 \\ 8 & 6 \\ 0 & 2 \end{pmatrix}$.

3. Work out:

    (a) $\begin{pmatrix} 2 & 1 & 0 \\ 3 & 2 & 1 \end{pmatrix}\begin{pmatrix} 2 & 1 \\ 3 & 0 \\ 4 & 1 \end{pmatrix}$;

    (b) $\begin{pmatrix} 3 & 4 \\ 5 & 6 \end{pmatrix}\begin{pmatrix} 2 & 3 \\ 4 & 5 \end{pmatrix}$;

    (c) $\begin{pmatrix} 4 & 5 & 6 & 0 \\ 0 & 1 & 2 & 3 \end{pmatrix}\begin{pmatrix} 1 & 3 \\ 2 & 2 \\ 3 & 1 \\ 4 & 0 \end{pmatrix}$;

    (d) $\begin{pmatrix} 2 & 3 \\ 1 & 5 \\ 6 & 0 \end{pmatrix}\begin{pmatrix} 5 & 0 & 6 \\ 2 & 7 & 3 \end{pmatrix}$;

    (e) $\begin{pmatrix} 3 & 0 & 1 \\ 0 & 5 & 4 \\ 2 & 1 & 5 \end{pmatrix}\begin{pmatrix} 6 & 0 & 1 \\ 2 & 3 & 4 \\ 3 & 2 & 1 \end{pmatrix}$;

    (f) $\begin{pmatrix} 1 & 2 \\ 0 & 5 \\ 3 & 1 \\ 4 & 2 \end{pmatrix}\begin{pmatrix} 6 & 3 & 7 \\ 0 & 4 & 1 \end{pmatrix}$.

Matrices

4. Let

$$A = \begin{pmatrix} 3 & 2 & 1 \\ 4 & 2 & 3 \\ 5 & 6 & 7 \end{pmatrix}, \quad B = \begin{pmatrix} 2 & 1 & 1 \\ 3 & 1 & 2 \end{pmatrix},$$

$$C = \begin{pmatrix} 4 & 7 \\ 5 & 8 \\ 6 & 9 \end{pmatrix} \quad \text{and} \quad D = \begin{pmatrix} 5 & 4 & 3 \\ 2 & 1 & 0 \end{pmatrix}.$$

(a) Write down the orders of the matrices **A**, **B**, **C** and **D**.
(b) We can find **BC**. Why? Find **BC**.
(c) Can we find **DA**? Give a reason for your answer. Find **DA** if possible.
(d) Can we find **AB**? Give a reason for your answer. Find **AB** if possible.

5. (a) When a 3 by 4 matrix is multiplied by a 4 by 2 matrix, what is the order of the answer? (Remember the domino pattern.)
(b) When a 2 by 5 matrix is multiplied by a 5 by 7 matrix, what is the order of the answer?

6. Let

$$E = \begin{pmatrix} 3 & 5 \\ 2 & 6 \\ 1 & 9 \end{pmatrix}, \quad F = \begin{pmatrix} 5 & 6 & 3 & 8 \\ 1 & 2 & 1 & 2 \\ 4 & 0 & 7 & 5 \end{pmatrix},$$

$$G = \begin{pmatrix} 2 & 7 & 0 & 9 \\ 1 & 3 & 8 & 6 \end{pmatrix} \quad \text{and} \quad H = \begin{pmatrix} 2 \\ 9 \\ 5 \end{pmatrix}.$$

(a) What is the order of matrix **X** if **EX** = **F**?
(b) What is the order of matrix **Y** if **YF** = **G**?
(c) What is the order of matrix **Z** if **EZ** = **H**?

7. Each week the orders of three houses from a baker are as follows:

	White loaves	Wholemeal loaves	Hovis
No. 1	1	1	0
No. 2	0	1	2
No. 3	1	2	1

White bread costs 7p a loaf, wholemeal 6p and Hovis 5p. Write the prices as a *column matrix*.

Multiply two matrices together to find the total bread bill for each house.

Matrix multiplication

8. 'Bildit' is a constructional toy with standard parts called flats, pillars, blocks, rods and caps. It is boxed in sets, numbered 1 to 3. Set 1 has 1 flat, 4 pillars, 8 blocks, 14 rods and 2 caps. Set 2 has 2 flats, 10 pillars, 12 blocks, 30 rods and 4 caps. Set 3 has 4 flats, 24 pillars, 30 blocks, 60 rods and 10 caps. Tabulate this information in a 3 by 5 matrix.

   Flats cost 5p each, pillars 4p, blocks 1p, rods 2p, and caps 3p. Write a suitable matrix for these prices and multiply two matrices together to find the cost of the various sets.

9. A factory produces three types of portable radio set called Audio 1, Audio 2 and Audio 3. Audio 1 contains 1 transistor, 10 resistors and 5 capacitors, Audio 2 contains 2 transistors, 18 resistors and 7 capacitors and Audio 3 contains 3 transistors, 24 resistors and 10 capacitors. Arrange this information in a matrix with sets in columns and parts in rows.

   Find the factory's weekly consumption of transistors, resistors and capacitors if its weekly output of sets is 100 of Audio 1, 250 of Audio 2 and 80 of Audio 3.

10. $A = \begin{pmatrix} 2 & 0 \\ 0 & 2 \end{pmatrix}$, $B = \begin{pmatrix} 0 & 2 \\ 1 & 1 \end{pmatrix}$ and $C = \begin{pmatrix} 1 & 2 \\ 3 & 1 \end{pmatrix}$.

    Work out: (a) **AB**; (b) **BA**; (c) **BC**; (d) **CB**. What do your answers tell you about matrix multiplication?

11. $P = \begin{pmatrix} 2 & 0 & 3 \\ 1 & 3 & 0 \end{pmatrix}$, $Q = \begin{pmatrix} 5 & 4 \\ 1 & 0 \\ 1 & 3 \end{pmatrix}$ and $R = \begin{pmatrix} 2 & 0 \\ 1 & 4 \end{pmatrix}$.

    Work out: (a) **(PQ)R**; (b) **P(QR)**. Make sure you write the matrices in the right order! Is there any confusion if we write **PQR**?

12. Write down any three matrices **A**, **B** and **C** each of order 2 by 2. Work out: (a) **(AB)C**; (b) **A(BC)**. Do you think that matrix multiplication is associative?

4. Experiments

1. EXPERIMENTS

This chapter is concerned with experiments and we start by doing one together as a class.

You will each need a die, which can be made by plaiting if you wish. Throw this die for several minutes, noting each time what number comes up. Decide how you can record your results. Afterwards, count how many times each number came up.

Now pool your results, and work out how many times each number came up for the class as a whole. Discuss this result.

Here are some more experiments, to be done by yourself or in small groups. You must repeat each experiment at least 100 times, and, if possible, more than this.

1. Toss a coin, noting each time whether a head or a tail comes up. When you have tossed the coin enough times, count how many heads and how many tails you obtained. Is this result what you would expect?

Experiments

2. Toss two coins at the same time. You will either get two heads, two tails or one of each. Note which of these you obtain each time. At the end of the experiment draw a bar chart to illustrate your results. Do you find the shape of this bar chart surprising? Try to work out why it is this shape.

3. Toss three coins. There are four possible ways in which they can land. What are these ways? Record your results as you did in the last experiment, and again draw a bar chart to show your results.

   In what ways is this chart similar to the one you got in the last experiment and in what ways is it different? Can you explain this?

4. Throw two dice and add the numbers together. Before you start, work out what scores are possible. Record your results carefully while you are doing the experiment and afterwards count how many times each score occurred. Draw a bar chart to illustrate this result. The beginning of the bar chart might be something like Figure 1.

   Try to explain the shape of your bar chart.

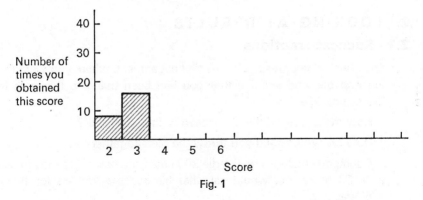

Fig. 1

5. For this experiment you need a nail maze. This is a piece of wood with nails arranged in the pattern shown in Figure 2.

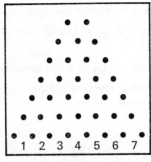

Fig. 2

49

Experiments 4

Arrange the board so that it slopes gently, with the numbers at the bottom. Place a marble at the top of the maze and let it roll downwards. Note in which hole the marble finishes.

Repeat the experiment a number of times. Draw a bar chart to show how many times the marble landed in each hole and try to explain the shape of your chart.

6. Put 3 red, 2 yellow and 5 blue counters into a box and, without looking, draw one out and note its colour. Put it back in the box, shake the box, and repeat the experiment many times.

How many times did you draw out a counter of each colour? Suppose you had not known how many counters of each colour there were in the box. Could you have worked this out from your results?

Try this now. Get a friend to put 10 counters in a box, but without telling you how many there are of each colour. Repeat Experiment 6 with these counters and work out how many of each colour you think there are. Look at the counters and see if you are right.

2. LOOKING AT RESULTS
2.1 Success fractions

Look back at your results for the first experiment, where you were throwing one die, and imagine that you had been trying to throw as many '5's as possible.

How many times did you succeed in throwing a '5'?

How many times did you throw the die altogether?

If someone had thrown the die 300 times and had succeeded in getting a '5' 53 times, we would say that his *success fraction* for throwing a '5' was $\frac{53}{300}$.

What was *your* success fraction for throwing a '5'?

What success fraction for throwing a '5' did each of your friends get?

What would you expect this success fraction to be?

Are your results and your friends' results the same as this expected value or near to it?

Why don't you all get exactly the same success fraction?

Exercise A contains some questions where you will work out some more success fractions.

Looking at results

Exercise A

In Questions 1 to 5, use the results from the experiments you have done already. For each of the other questions you must do a new experiment.

1 What was your success fraction for getting a head when you tossed 1 coin?

2 When you tossed 2 coins, what were your success fractions for getting:
(a) 1 head and 1 tail; (b) 2 tails?

3 When you tossed 3 coins, what were your success fractions for getting:
(a) 3 heads; (b) 2 heads and 1 tail; (c) 2 tails and 1 head; (d) 3 tails?

4 When you threw 2 dice, what were your success fractions for scoring:
(a) 7; (b) 11; (c) 10; (d) 3?

5 In Experiment 6, what were your success fractions for pulling out:
(a) a blue counter; (b) a red counter?

6 Get a friend to take 13 cards out of a full pack of 52 cards (no jokers), and hold them so that you cannot see what they are. Pick out a card, look at it, note its suit and replace it. Repeat this about 60 times. What were your success fractions for pulling out:
(a) a heart; (b) a spade; (c) a club; (d) a diamond?

Look at the cards. How many of each suit were there? Does this make sense when compared with your success fractions?

7 Write the numbers 1 to 10 on pieces of paper. Fold up the pieces and put them in a box. Draw out a piece of paper, note its number and replace it in the box. Repeat this a suitable number of times. What were your success fractions for pulling out:
(a) 6; (b) an even number; (c) a prime number?

8 Write down the birthday month of each member of the class. What fraction of pupils in the class have birthdays in:
(a) June; (b) April; (c) November; (d) October?

9 Write down the day of the month on which the birthday of each person in the class occurs. What fraction of the pupils in the class has a birthday on:
(a) the 22nd; (b) the 6th; (c) the 3rd;
(d) between the 1st and the 10th;
(e) between the 20th and the 31st?

10 Find what fraction of pupils in your class is left-handed.

Experiments 4

11 If you threw 3 dice and added up the numbers, what possible scores could you get? Throw 3 dice a large number of times (at least 200). What were your success fractions for scoring:
(a) 3;   (b) 12;   (c) 9;   (d) 11;
(e) 19;   (f) 17;   (g) 6 or 7; (h) an odd number?

What score is most likely to come up?

2.2 Projects

1. Experiments like the ones you have done might help if your were running a fete and wished to make money out of stalls on which people threw dice, rolled coins, etc.

   Suppose you charged 1p a time for people to throw two dice, and gave a prize for each double (2 ones, 2 twos, etc.) and a special prize for a double six. By looking at your results to Experiment 4, try to work out how much the prizes could be if you wanted to give back about half your takings in prizes.

   Try to think of other stalls of this kind which you could run, and work out the prizes.

2. In Question 8 of Exercise A you found the fraction of people in your class with birthdays in a particular month. Suppose, in your class, the fraction of birthdays in November was $\frac{2}{33}$. Would you expect that the fraction of people in the whole country with birthdays in November would be equal to $\frac{2}{33}$?

   If not, work out what you would expect this fraction to be and explain your method.

3. Suppose a firm manufacturing potato peelers wished to know what fraction of its output should be left-handed potato peelers. You have found the fraction of left-handed people in your class. Would this be an accurate enough fraction for the firm to use?

   In the birthday month question it was easy to work out a sensible answer for the number of people born in November ($\frac{1}{12}$ or, more accurately, $\frac{30}{365}$). Is there a simple method like this for working out the fraction of left-handed people? If not, why not?

4. When information is required which cannot reasonably be worked out by counting, a survey is taken. Why do the best surveys question a large number of people?

   Think of some information about people in your school that it would be useful to have and conduct a survey. You will have to question as many people as possible, and about the same number from each year in the school.

Interlude

THINGS ARE NOT WHAT THEY SEEM

(a) Figure 1 shows three well known sayings. What do they say?

Fig. 1

Read them again—*carefully*.

Did you read them correctly the first time or did you read what you *expected* to get?

Diagrams can be misleading so look at them carefully and critically.

(b) Look at Figure 2. It shows the vitamin and mineral content of 1 kg of 'Doggo' compared with that of 1 kg of shin of beef. Is this a fair way of showing this information? Give reasons for your answer.

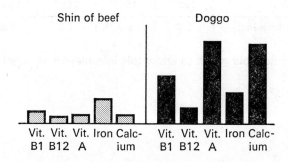

Fig. 2

Interlude

(c) The graph in Figure 3 shows the swing to 'Wosh'—the wonder detergent. Criticize it.

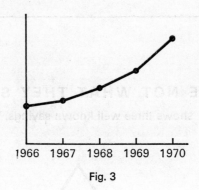

Fig. 3

(d) This table shows the annual sales of 'Wosh'.

Year	1966	1967	1968	1969	1970
Amount sold in tonnes	220	230	250	270	310

Draw two graphs to illustrate this information:
  (i) with the scale up the page marked from 0 to 500 tonnes,
  (ii) with the scale up the page marked from 200 to 400 tonnes.

Which gives the most accurate representation?

If you were a salesman for this particular product, which graph would you prefer to show to prospective customers? Why?

(e) The following table gives information concerning the sales of the 'Henbury Times'.

Year	1966	1967	1968	1969	1970
Sales in thousands	5·8	6·3	7·3	8·8	10·9

Draw a suitable graph to show this information as clearly as possible.

Things are not what they seem

(f) The graph in Figure 4 shows the percentage of wins of a school basket ball team over the last four years.

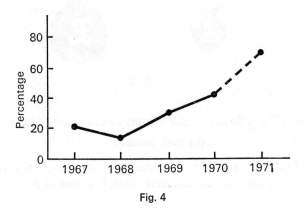

Fig. 4

The dotted line shows the 'predicted percentage wins' for the 1971 season. Do you think this would be reasonable? Give reasons for your answer.

(g)

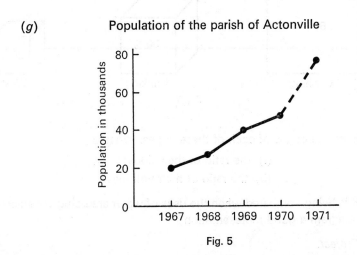

Fig. 5

'The population of Actonville for 1971 will probably be about 75 000'. Criticize this statement.

Interlude

(*h*) Figure 6 illustrates the company chairman's statement 'Profit this year is double that of last year'. Is the figure a fair representation?

Fig. 6

What is the ratio of: (i) the heights of the money bags;
(ii) their areas?

(*i*) Statistical pictures or 'pictograms' can also be very misleading. Figure 7 shows some information about the uses of water.

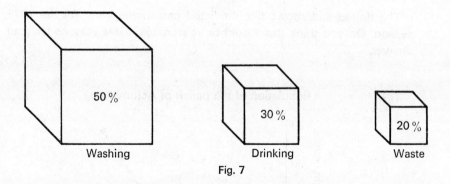

Fig. 7

Measure the edges of each of these cubes. What is
(i) the ratio of their sides;
(ii) the ratio of their volumes?

What is being compared when the uses of water are being considered? Is this diagram a good representation?

(*j*) *Project.*

Make a collection of statistical representations from newspapers and magazines. Look at each one carefully and discuss whether or not it is misleading.

Revision exercises

Computation 1
1. $81108 \div 2 \cdot 7$.
2. $2 \cdot 1 \times 5 \div 7$.
3. $41 \cdot 8 + 8 \cdot 67 - 0 \cdot 83 + 13$.
4. $12 \cdot 5 \times 37 \cdot 5$.
5. Find the mean of 48, 52, 49, 46, 50.
6. $\begin{pmatrix} 0 & 3 \\ 4 & -2 \end{pmatrix} + \begin{pmatrix} 6 & -7 \\ -1 & 0 \end{pmatrix}$.

Exercise A
1. A triangle has sides of 4 cm, 5 cm and 6 cm. Is it right-angled?
2. What is 5% of 300?
3. Write $2p + 2p + 3p - p$ in shorter form.
4. Is the letter A traversable?
5. Find the value of $4(a-b)$ when $a = 3$ and $b = {}^-5$.
6. $A = \{2, 3, 5, 7\}$, $B = \{2, 4, 6, 8, 10\}$. What is $A \cap B$?
7. One angle of an isosceles triangle is 110°. What are the sizes of the others?
8. Carry out the matrix multiplication
$$\begin{pmatrix} 2 & 1 \\ 0 & 4 \end{pmatrix} \begin{pmatrix} -1 & 0 \\ 3 & -1 \end{pmatrix}.$$

Exercise B
1. What is the length of the diagonal of a rectangle whose sides are 6 cm and 8 cm?
2. Three angles of a quadrilateral are 48°, 87°, 144°. Find the fourth.
3. Convert 65_{ten} to base eight.

Revision exercises

4 Draw a network described by the following matrix:

$$\begin{pmatrix} 2 & 1 & 0 & 1 \\ 1 & 0 & 3 & 1 \\ 0 & 3 & 0 & 2 \\ 1 & 1 & 2 & 4 \end{pmatrix}.$$

5 $A = \{1, 2, 3, 4, 5\}$, $B = \{2, 4, 6, 8, 10\}$. How many members has $A \cup B$?

6 Where is the image of the point $(^-3, 1)$ after reflection in the $y = 0$ axis?

7 What linear relation do the following points satisfy?

(0, 1), (1, 4), (2, 7), (3, 10).

8 A firm employs 240 people. Two-thirds of them are men. How many are women?

Exercise C

1 Figure 1 shows an octagon. By using Pythagoras's rule, find the area of the red square.

The area of the shaded square is twice the area of the red square. Find the dimensions of the shaded square. Hence find the area of the whole octagon.

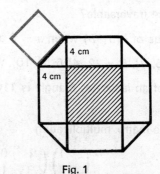

Fig. 1

2 List the members of $A \cap B$ for each of the following pairs of sets:
(a) $A = \{3, 5, 7, 9, 11\}$, $B = \{$odd numbers between 2 and 10$\}$;
(b) $A = \{$the vowels$\}$, $B = \{$the first five letters of the alphabet$\}$;
(c) $A = \{$letters in your surname$\}$, $B = \{$letters in your first name$\}$;
(d) $A = \{$pupils in your class who wear glasses$\}$,
  $B = \{$pupils in your class who do not wear glasses$\}$.

Revision exercises

3 The four houses of a co-educational school are named after the planets Mars, Pluto, Saturn and Jupiter. The way in which first form boys and girls were allocated to these houses is described by the following matrices:

$$\text{Form A} \quad \begin{array}{c} \text{M P S J} \\ \text{boys} \begin{pmatrix} 3 & 2 & 5 & 5 \\ 4 & 6 & 2 & 4 \end{pmatrix} \end{array} \quad \text{Form B} \quad \begin{array}{c} \text{M P S J} \\ \text{boys} \begin{pmatrix} 5 & 4 & 2 & 1 \\ 4 & 3 & 5 & 5 \end{pmatrix} \end{array} \quad \text{Form C} \quad \begin{array}{c} \text{M P S J} \\ \text{boys} \begin{pmatrix} 3 & 6 & 3 & 4 \\ 3 & 2 & 6 & 5 \end{pmatrix} \end{array}$$

(a) Calculate from these matrices a 2 by 4 matrix describing the allocation of boys and girls to houses in the first form as a whole.
(b) Which house has the largest number of
  (i) boys; (ii) girls; (iii) first formers?

4 If you threw three tetrahedral dice, each numbered from 1 to 4, what possible total scores could you get?

5 Copy Figure 2 onto squared paper and enlarge it by a scale factor of $1\frac{1}{2}$ using the star as the centre of enlargement. Repeat for scale factors of $\frac{1}{2}$ and $-\frac{1}{2}$.

Calculate the area of each kite and hence find a relation between the area of the original kite and each enlargement. Are the relations what you would have expected?

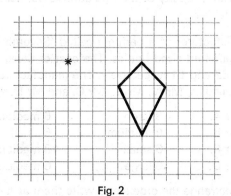

Fig. 2

6 Sketch the following figures and on each mark clearly any lines of symmetry and centres of rotational symmetry.
  (a) Square; (b) rectangle; (c) parallelogram; (d) rhombus;
  (e) kite; (f) isosceles triangle; (g) equilateral triangle.

Revision exercises

Exercise D

1. A triangle has vertices at $(-2, -1)$, $(4, -1)$ and $(1, 3)$. What kind of triangle is it? Find the lengths of its sides, using Pythagoras's rule where necessary.

2. If $\emptyset$ is the empty set, $A = \{a\}$, $B = \{b\}$ and $P = \{a, b\}$, copy and complete the following table. ($\emptyset \cap B = \emptyset$ and $A \cap P = A$ have already been entered for you.)

		Second member			
$\cap$		$\emptyset$	A	B	P
First member	$\emptyset$			$\emptyset$	
	A				A
	B				
	P				

(a) Is $\{\emptyset, A, B, P\}$ closed under intersection?
(b) Does $\{\emptyset, A, B, P\}$ contain an identity? If so, what is it?
(c) Sketch in lightly two lines running diagonally across the table. Which one could be called a line of symmetry?
(d) What property of the operation is connected with the symmetry of the table?

3. A manufacturer of a perfume called 'Seventh Heaven' markets it in three different sized bottles. The cost of these bottles to the chemist is £3, £2 and £1 respectively. The manufacturer receives orders from four chemists, Mr A., Mr B., Mr C. and Mr D. as follows:

Mr A.	3 large	6 medium	10 small
Mr B.	0 medium	6 small	12 large
Mr C.	6 small	18 medium	6 large
Mr D.	0 large	6 medium	24 small

(a) Rearrange the orders and write them as a 4 by 3 matrix.
(b) Write the costs as a column matrix.
(c) Multiply the two matrices together to find the total bill for each chemist.

4. 3 red, 3 blue and 3 white counters are placed in a hat. Two of these counters are then picked from the hat without looking. List all the possible results.

Revision exercises

5 Figure 3 (not drawn accurately) represents the net of a non-regular tetrahedron with one corner sliced off.

(a) State which lengths *must* be equal.

(b) What point will coincide with *J* when the polyhedron is formed?

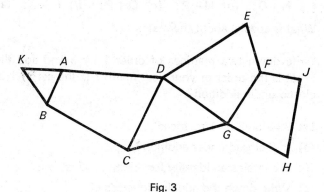

Fig. 3

6 Draw the graphs of $x - y = 6$ and $x = 1$. Show by shading, the region where $x - y < 6$ and $x > 1$.

MORE PRACTICE WITH MATRICES
Exercise E

1 Let
$$A = \begin{pmatrix} 3 & -2 \\ 1 & 6 \\ -4 & 0 \end{pmatrix} \text{ and } B = \begin{pmatrix} -3 & 1 \\ 2 & 7 \\ -2 & 4 \end{pmatrix}.$$

Find

(a) **A + B**;   (b) **A − B**;   (c) **B − A**;

(d) **3A**;   (e) **B + 3A**;   (f) **3A − B**.

2 Let
$$X = \begin{pmatrix} 1 & 0 & -8 \\ -2 & 7 & 3 \end{pmatrix}, \quad Y = \begin{pmatrix} -3 & 5 & 2 \\ 2 & -7 & 4 \end{pmatrix} \text{ and } Z = \begin{pmatrix} 0 & -1 & 3 \\ 2 & 6 & -9 \end{pmatrix}.$$

Find:

(a) **X − Z**;   (b) **X + Y**;   (c) **X + Y + Z**;   (d) **X + Y − Z**.

Revision exercises

3. Let
$$N = \begin{pmatrix} 5 & 0 & 0 \\ 7 & 8 & -1 \\ -4 & 3 & 2 \end{pmatrix}, \quad O = \begin{pmatrix} 0 & 0 & 0 \\ 0 & 0 & 0 \\ 0 & 0 & 0 \end{pmatrix} \quad \text{and} \quad P = \begin{pmatrix} -5 & 3 & 9 \\ 2 & 0 & -3 \\ -1 & -4 & 7 \end{pmatrix}.$$

   Find:

   (a) $N+O$; (b) $N+P$; (c) $O+P$; (d) $P+O$; (e) $O+N$.

   What is special about the matrix O?

4. Write down two matrices of order 1 by 3 and add them together. What is the order of your new matrix? Is the set of 1 by 3 matrices closed under addition?

5. Let $S = \{$all 2 by 1 matrices$\}$.
   (a) Is S closed under addition?
   (b) S contains an identity for addition. What is it?
   (c) Write down the additive inverses of

   (i) $\begin{pmatrix} 0 \\ 1 \end{pmatrix}$; (ii) $\begin{pmatrix} 4 \\ -7 \end{pmatrix}$; (iii) $\begin{pmatrix} -2 \\ 0 \end{pmatrix}$.

6. Let $M = \{$all 2 by 2 matrices$\}$.
   (a) Is M closed under addition?
   (b) Does M contain an identity for addition? If so, what is it?
   (c) Write down the additive inverses of

   (i) $\begin{pmatrix} 2 & 0 \\ 0 & 2 \end{pmatrix}$; (ii) $\begin{pmatrix} -3 & 5 \\ 0 & 7 \end{pmatrix}$; (iii) $\begin{pmatrix} 2 & -1 \\ 4 & -3 \end{pmatrix}$.

Exercise F

1. Work out:

   (a) $\begin{pmatrix} 1 & 0 \\ 0 & 1 \end{pmatrix}\begin{pmatrix} 7 \\ 3 \end{pmatrix}$;

   (b) $(5 \ -8)\begin{pmatrix} 1 & 0 \\ 0 & 1 \end{pmatrix}$;

   (c) $\begin{pmatrix} 1 & 0 \\ 0 & 1 \end{pmatrix}\begin{pmatrix} 2 & -3 \\ -4 & 5 \end{pmatrix}$;

   (d) $\begin{pmatrix} 2 & -3 \\ -4 & 5 \end{pmatrix}\begin{pmatrix} 1 & 0 \\ 0 & 1 \end{pmatrix}$.

   What do you notice? What is special about the matrix

   $\begin{pmatrix} 1 & 0 \\ 0 & 1 \end{pmatrix}$?

More practice with matrices

2 Work out:

(a) $\begin{pmatrix} 1 & 0 & 0 \\ 0 & 1 & 0 \\ 0 & 0 & 1 \end{pmatrix} \begin{pmatrix} 2 \\ 3 \\ 4 \end{pmatrix}$; (b) $\begin{pmatrix} -5 & 3 & 2 \\ 2 & -1 & 1 \end{pmatrix} \begin{pmatrix} 1 & 0 & 0 \\ 0 & 1 & 0 \\ 0 & 0 & 1 \end{pmatrix}$.

See if you can write down the answer to these without working them out:

(c) $\begin{pmatrix} 1 & 0 & 0 \\ 0 & 1 & 0 \\ 0 & 0 & 1 \end{pmatrix} \begin{pmatrix} -5 & 6 & 3 \\ 4 & 4 & -5 \\ 7 & -7 & 1 \end{pmatrix}$; (d) $\begin{pmatrix} 4 & 5 & 6 \\ -2 & 7 & -8 \\ 5 & 9 & -3 \\ 6 & -4 & 4 \end{pmatrix} \begin{pmatrix} 1 & 0 & 0 \\ 0 & 1 & 0 \\ 0 & 0 & 1 \end{pmatrix}$.

3 Copy this multiplication and see if you can fill in the blanks:

$$\begin{pmatrix} 5 & 6 & -7 & 8 \\ 9 & -2 & 4 & 1 \\ -3 & 5 & 6 & 7 \\ 2 & 1 & 0 & -3 \end{pmatrix} \begin{pmatrix} - & - & - & - \\ - & - & - & - \\ - & - & - & - \\ - & - & - & - \end{pmatrix} = \begin{pmatrix} 5 & 6 & -7 & 8 \\ 9 & -2 & 4 & 1 \\ -3 & 5 & 6 & 7 \\ 2 & 1 & 0 & -3 \end{pmatrix}.$$

4 Work out:

(a) $\begin{pmatrix} 3 & -5 \\ 4 & 2 \end{pmatrix} \begin{pmatrix} 2 & 0 \\ 1 & 4 \end{pmatrix}$; (b) $\begin{pmatrix} 2 & -1 \\ 3 & 2 \end{pmatrix} \begin{pmatrix} 4 & 1 \\ 3 & -2 \end{pmatrix}$;

(c) $\begin{pmatrix} 5 & -1 & 3 \\ 0 & 2 & -1 \end{pmatrix} \begin{pmatrix} 5 & 2 & 0 \\ 0 & -1 & 4 \\ 1 & 3 & -7 \end{pmatrix}$; (d) $\begin{pmatrix} 3 & 1 \\ -1 & 2 \\ 0 & 2 \end{pmatrix} \begin{pmatrix} 0 & 7 \\ -4 & 1 \end{pmatrix}$;

(e) $\begin{pmatrix} 7 & 3 \\ -1 & 5 \end{pmatrix} \begin{pmatrix} 4 & 0 & -1 & 1 \\ 0 & -2 & 1 & -3 \end{pmatrix}$;

(f) $\begin{pmatrix} 2 & 1 & -5 \end{pmatrix} \begin{pmatrix} 7 & 2 \\ 1 & 4 \\ 0 & -1 \end{pmatrix} \begin{pmatrix} 3 & 0 \\ 0 & -2 \end{pmatrix}$.

5 We write **AA** as A^2, AA^2 as A^3, AA^3 as A^4 and so on.
If
$$A = \begin{pmatrix} 3 & 1 \\ -2 & 0 \end{pmatrix},$$
find (a) A^2; (b) A^3; (c) A^4.

Revision exercises

6   If
$$B = \begin{pmatrix} 0 & 1 \\ -1 & 0 \end{pmatrix},$$
find (a) B^2; (b) B^3; (c) B^4.

What do you notice about your last answer?
Write down the answers to these without working them out:
(d) B^5;   (e) B^6;   (f) B^7;   (g) B^8.

7   Let
$$P = \begin{pmatrix} 5 & 3 \\ 3 & 2 \end{pmatrix} \text{ and } Q = \begin{pmatrix} 2 & -3 \\ -3 & 5 \end{pmatrix}.$$

(a) Work out: (i) PQ; (ii) QP. What do you notice?
(b) Write down the inverse of P for matrix multiplication.
(c) Write down the multiplicative inverse of Q.

8   Let
$$M = \begin{pmatrix} 1 & 0 \\ 0 & -1 \end{pmatrix} \text{ and } N = \begin{pmatrix} -1 & 0 \\ 0 & 1 \end{pmatrix}.$$

(a) Find (i) M^2; (ii) N^2. What do you notice?
(b) What is special about the inverse properties of matrices M and N?
(c) Write down the answers to these without working them out:

(i) M^3;   (ii) N^4.

9   Let
$$F = \begin{pmatrix} 0 & -1 \\ 1 & 0 \end{pmatrix}, \quad G = \begin{pmatrix} 0 & 1 \\ -1 & 0 \end{pmatrix},$$
$$H = \begin{pmatrix} -1 & 0 \\ 0 & -1 \end{pmatrix} \text{ and } I = \begin{pmatrix} 1 & 0 \\ 0 & 1 \end{pmatrix}.$$

Copy and complete the following table for matrix multiplication:

Matrix multiplication	Right-hand matrix			
		F	G	H	I
Left-hand matrix	F	H			
	G				G
	H				
	I			H	

(a) Is {F, G, H, I} closed under matrix multiplication?
(b) Is there an identity? If so, what is it?
(c) What are the inverses, if any, of F, G, H, I?

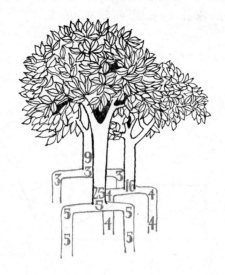

5. Square roots

1. SQUARE ROOTS BY CALCULATION

1.1 What we mean by a square root

In Chapter 1 we met a situation like this:

Take a right-angled triangle with measurements as in Figure 1.

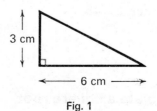

Fig. 1

Square roots

Draw the squares on its sides as in Figure 2.

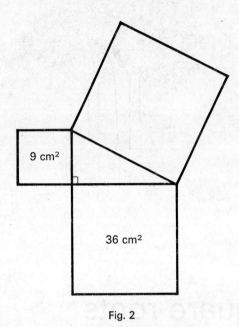

Fig. 2

Notice that although we do *not* know the length of the side of the largest square, we do know its area.

Pythagoras's rule gives the area of the largest square as

$$9 + 36 = 45 \text{ cm}^2.$$

The length (in cm) of the side of the largest square is called the *square root* of 45, and is written as $\sqrt{45}$.

1.2 How to calculate a square root

We can see that

$\sqrt{45}$ is more than 6 because $6 \times 6 = 36$,

and less than 7 because $7 \times 7 = 49$.

Square roots by calculation

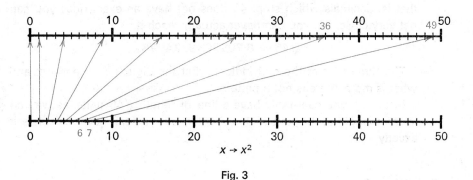

Fig. 3

That is, $\quad 6 < \sqrt{45} < 7,$

or in words, the square root of 45 is between 6 and 7.

We must therefore look for a number which is six point something, and which, when multiplied by itself, gives us 45:

$$\square \times \square = 45.$$

Try 6·7.

This gives the area of the square as $6·7 \times 6·7 = 44·89$ cm² which is too small.

Try 6·8.

This gives the area of the square as $6·8 \times 6·8 = 46·24$ cm² which is too big.

Now we can say: $6·7 < \sqrt{45} < 6·8.$

Try 6·75.

This number squared produces $6·75 \times 6·75 = 45·5625$ which is a little too big, but we are now getting very close.

Between which two numbers can we now say that the square root of 45 lies?

What would you do if you wanted to get closer still?

Use either long multiplication of decimals or desk calculating machines to continue the process just started and see if you can find the exact square root of 45.

Problems like these were of great concern to the early mathematicians. They found that very few numbers have square roots which are *exact*,

Square roots

that is, decimals which stop. 45 does not have an exact one; you can get very close, but you can never actually reach it.

$$\sqrt{45} = 6{\cdot}708\,203\,932\,499\,3\ldots$$

The three dots at the end indicate that the digits go on and on, and what is more, there is not a pattern.

In theory, you can easily have a line of length $\sqrt{45}$ units, as we had in the triangle at the start of this chapter, but you can never calculate it exactly.

Exercise A

1. Write down the first ten square numbers. Beneath each one write its square root.

2. The following numbers have exact square roots. What are they?
   (a) 144;   (b) 400;   (c) 169;   (d) 900;   (e) 1·44;   (f) 0·25.

3. Find the length of a side of each of the squares in Figure 4.

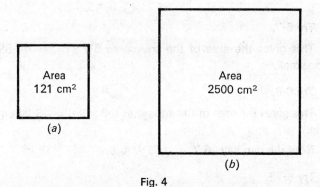

Fig. 4

4. Find two *whole* numbers between which the square root of each of the following numbers lies. Try to get them as close together as you can.
   (a) 70;   (b) 150;   (c) 200;   (d) 600;   (e) 1200.

5. Show by multiplication that the following statement is correct:
   $$3{\cdot}60 < \sqrt{13} < 3{\cdot}61.$$

6. Between which two whole numbers does $\sqrt{40}$ lie? By considering numbers other than whole numbers, find a more accurate estimate which is correct to 1 decimal place.

Square roots by calculation

7 Estimate the value of the following numbers correct to 1 decimal place:
   (a) $\sqrt{8}$;   (b) $\sqrt{18}$;   (c) $\sqrt{33}$;   (d) $\sqrt{75}$.

8 Which is the larger? Test by squaring.
   (a) $\sqrt{5}$ or $2\frac{2}{9}$;   (b) 2·6 or $\sqrt{7}$.

9 Complete the table and use it to draw your own graph like the one in Figure 5, but double both scales for greater accuracy.

x		1		4	$6\frac{1}{4}$	
$\sqrt{x}$	$\frac{1}{2}$		$1\frac{1}{2}$			3

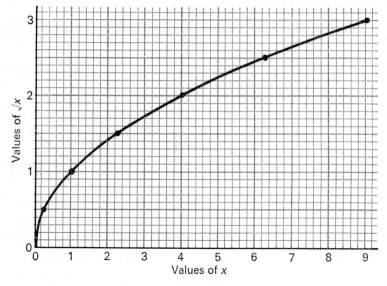

Fig. 5

Use your graph to give values for:
(a) $\sqrt{3}$; (b) $\sqrt{5}$; (c) $\sqrt{8}$; (d) $\sqrt{4\cdot5}$; (e) $\sqrt{2\cdot6}$;
(f) 2·2 × 2·2; (g) 1·4 squared; (h) $(1\cdot8)^2$.

10 Find the value of
    (a) $\sqrt{81}$;   (b) $\sqrt{9^2}$;   (c) $\sqrt{15^2}$;   (d) $\sqrt{(2\cdot3)^2}$.

11 What is the area of a square which has a side of length:
    (a) $\sqrt{11}$ cm;   (b) $\sqrt{30}$ cm;   (c) $\sqrt{49}$ cm?

Square roots

2. SQUARE ROOTS BY DRAWING

We have seen that a length of $\sqrt{45}$ units can be found by accurate drawing. Other lengths can also be found in this way. For example, accurate drawing of the triangles in Figure 6 would give lengths of $\sqrt{2}$, $\sqrt{5}$, $\sqrt{17}$ and $\sqrt{13}$ units.

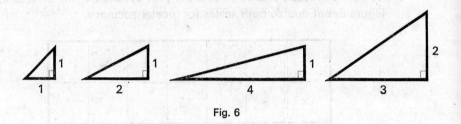

Fig. 6

Use Pythagoras's rule to check this.
Which square roots would be given by drawing the triangles in Figure 7?

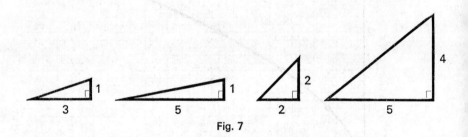

Fig. 7

Draw two of these triangles for yourself and estimate the value of the square roots.

2.1 Square root patterns

Here are two square root patterns which will produce the square roots of all the whole numbers in turn. Make an accurate drawing of each pattern and try to see how and why each works.

(a) Draw a pair of parallel lines 1 unit apart. Put a compass point at O and draw arc a, then draw the perpendicular p. Keep the compass point at O and draw arc b, then draw perpendicular q. Continue this process for as long as you wish.

Square roots by drawing

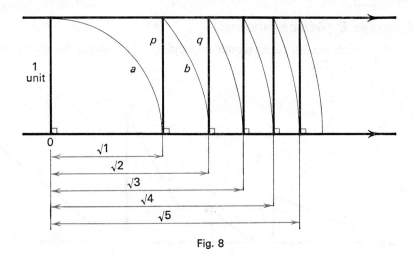

Fig. 8

(*b*) Start by drawing the small shaded triangle which has two sides of length 1 unit enclosing a right-angle. Be careful to place your first triangle so that the pattern does not run off the page. Continue the snail pattern for as long as you like.

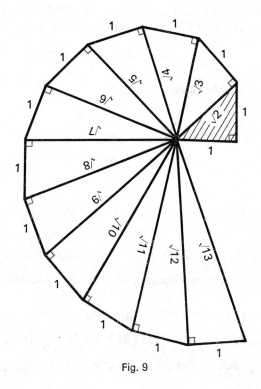

Fig. 9

Square roots

Exercise B (Miscellaneous)

1. Which square roots would be obtained by making the drawings in Figure 10?

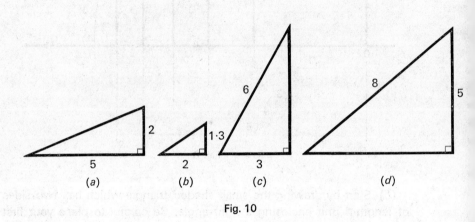

Fig. 10

2. Basil Brayne said that $\sqrt{25} = 5$, so $\sqrt{250} = 50$. What is wrong with this statement? Find two whole numbers, *a* and *b* such that

$$a < \sqrt{250} < b.$$

3. Copy and complete this pattern:

$$2^2 = 4$$
$$20^2 = 400$$
$$200^2 =$$
$$2000^2 = \quad .$$

Comment on the way in which the following two sequences are built up:

(a) 2, 20, 200, 2000, ...; (b) 4, 400, 40000, ...

What can you say about $\sqrt{40}$, $\sqrt{4000}$, $\sqrt{400000}$?

4. State whether the following are true or false:

(a) $\sqrt{36} = 6$;   (b) $\sqrt{360} = 60$;   (c) $\sqrt{4000} = 200$;

(d) $\sqrt{4900} = 70$;   (e) $\sqrt{16000} = 400$;   (f) $\sqrt{12100} = 110$.

Square roots by drawing

5 (Use desk calculators for this question if you have them.) See how close you can get to the square root of 20. Put your results in a table like the one which has been started below:

x	x^2	Too big	Too small
4·5	20·25	✓	
4·4	19·36		✓
4·45			

6 Use the method of Question 5 to see how close you can get to the value of:

(a) $\sqrt{2}$; (b) $\sqrt{60}$; (c) $\sqrt{14·5}$.

7 On graph paper draw two straight lines 10 cm long and about 3 cm apart (see Figure 11). Join points on the two lines by arrows so that numbers on the upper line are joined to their squares on the lower line. For example,

$$(1·5)^2 = 2·25, \text{ and } (0·2)^2 = 0·04,$$

so 1·5 is joined to 2·25 and 0·2 to 0·04.

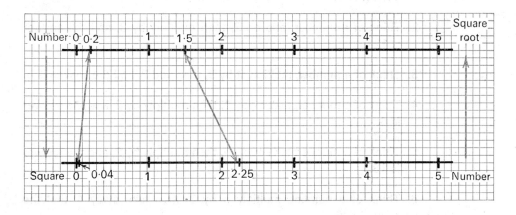

Fig. 11

Draw arrows from 0, 0·1, 0·2, 0·3, ..., 0·9, 1, 1·2, 1·5, 1·8, 2 and 2·2 on the upper line to corresponding points on the lower line. Comment on the pattern you find, particularly for the region between 0 and 1.

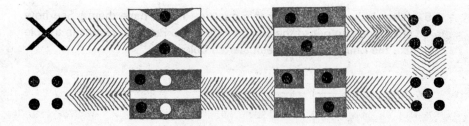

6. Solving equations

1. THINK OF A NUMBER

Here are some 'think of a number' questions. Try to work out what the number is each time.

1. Think of a number, multiply it by 2, then take away 3, and the answer is 5.

2. Think of a number, add on 2, then multiply by 3, and the answer is 12.

3. Think of a number, multiply it by 3, then add 2, and the answer is 8.

4. Think of a number, take away 1, then multiply by 2, and the answer is 4.

5. Think of a number, take away 5, then multiply by 2, and the answer is 18.

6. Think of a number, multiply it by 3, then take away 4, and the answer is 6.

You may have found the last two questions difficult. We are going to work out methods for answering questions like these.

First of all, we can save time by writing the questions more neatly—as equations.

Think of a number

In Question 1, if the number is called x we can write:
$$2x - 3 = 5.$$
In Question 2, we first add 2 and then multiply by 3, so we get:
$$(x+2) \times 3 = 12.$$
We usually write this as:
$$3(x+2) = 12,$$
which means exactly the same thing and looks neater.

Write the other four questions as equations.

2. SOLVING EQUATIONS

2.1 Graphical method

We shall now work out some ways of solving these equations; that is, of finding which value of x fits each equation.

In *Book D* you learned how to do this by drawing graphs. Figure 1 shows how to solve $2x - 3 = 5$ by drawing the graph of $y = 2x - 3$.

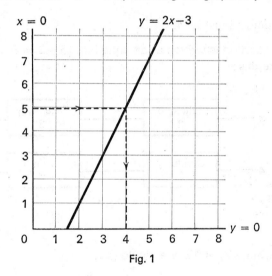

Fig. 1

This shows that if $2x - 3 = 5$, then $x = 4$.

Solve Question 6 and one of the other four equations by drawing graphs. (You need only use positive values of x and y for the graphs but in Question 5 both the scales must go up to 20.)

2.2 Flow diagrams

The graphical method is rather slow for solving equations and it does not work very well when the value of x is not a whole number.

What answer did you get for Question 6?
Are you sure that this is exactly right?

Solving equations 6

We could also solve these equations by drawing flow diagrams.

Figure 2 shows a flow diagram for Question 1:

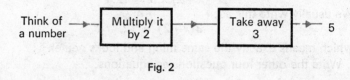

Fig. 2

In order to find the number, we reverse the diagram (see Figure 3).

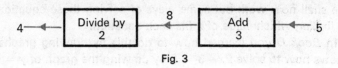

Fig. 3

The number must be 4.

If the question is written as an equation, $2x - 3 = 5$, the flow diagram looks like this:

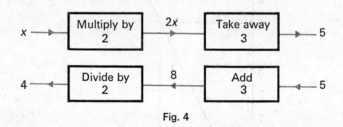

Fig. 4

So $x = 4$.

To solve $3(x+2) = 12$, we would get:

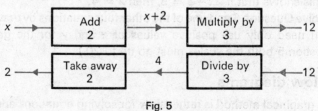

Fig. 5

So $x = 2$.

Solve the other four equations in this way.

Solving equations

Exercise A

Solve the following equations by drawing flow diagrams. When you have solved an equation check that you are right by making sure that your value of x does fit the equation.

1. $4x - 1 = 23$.
2. $3(x + 5) = 21$.
3. $3x + 7 = 13$.
4. $2(x - 3) = 12$.
5. $5(x + 2) = 25$.
6. $6x - 5 = 7$.

The flow diagram for Question 7 will start:

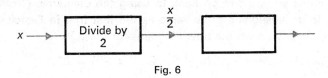

Fig. 6

7. $\frac{x}{2} + 3 = 11$.
8. $\frac{1}{2}x - 4 = 2$. ($\frac{1}{2}x$ may be written as $\frac{x}{2}$.)
9. $\frac{1}{3}(x - 4) = 2$.
10. $3(x - 2) = 4$.
11. $2(3x - 4) = 16$.
12. $3(\frac{1}{4}x + 2) = 12$.
13. $3x + 2 = 10$.
14. $5x - 1 = 11$.
15. $3x + 4 = 1$.
16. $2x + 6 = 2$.
17. $\frac{1}{2}(x + 5) = 2$.
18. $2(x + 7) = 4$.

2.3 Doing without flow diagrams

You will probably now find that you can solve some equations without actually drawing the flow diagrams.

We will try to do this for $2x - 3 = 5$. In the flow diagram (see Figure 4), we first multiplied by 2 and then took away 3. When we reversed the flow diagram we first added 3 to get $2x = 8$ and then divided by 2 to get $x = 4$.

Instead we could write:

If it is true that $2x - 3 = 5$
then $2x = 8$ must be true as well.
So $x = 4$ must be true.

Usually we leave out the explanations and just put:

$2x - 3 = 5$
$2x = 8$
$x = 4$.

Solving equations 6

Now we will solve $3(x+2) = 12$ in this way.
(*Think*: In the flow diagram we would add 2, and then multiply by 3. In reversing we would divide by 3 and then take away 2.)
So we get:
$$3(x+2) = 12$$
$$x+2 = 4$$
$$x = 2.$$

Exercise B

Try to solve the following equations without flow diagrams, but if you get muddled you can go back to using the diagrams. (Notice that the first four questions are the same as the first four in Exercise A, so you have already drawn the flow diagrams for these.)

1 $4x-1 = 23$.
2 $3(x+5) = 21$.
3 $3x+7 = 13$.
4 $2(x-3) = 12$.
5 $3x+2 = 11$.
6 $3(x-1) = 9$.
7 $5x+2 = 22$.
8 $2(x+3) = 12$.
9 $5x-1 = 19$.
10 $4(x-5) = 12$.
11 $\frac{1}{3}x+2 = 6$.
12 $\frac{1}{2}(x-2) = 3$.
13 $2(4x-1) = 18$.
14 $\frac{1}{2}(4x+1) = 6$.
15 $2(x+3) = 4$.
16 $2x+1 = 4$.
17 $3(\frac{x}{2}-1) = 12$.
18 $\frac{1}{2}(4x-3) = 6$.
19 $\frac{1}{3}(x+7) = 2$.
20 $2(\frac{1}{3}x+5) = 4$.

7. Probability

1. WHAT IS PROBABILITY?

Suppose that you are playing a game with a die and wish to throw a '6'.
The set of numbers which it is possible to obtain from throwing a die is

$$P = \{1, 2, 3, 4, 5, 6\}.$$

The set of numbers in which you are interested is

$$S = \{6\}.$$

If the die is a perfect cube (and is not loaded!) then we may reasonably suppose that we are *equally likely* to throw any one of the six numbers on it.

The set P contains *six* possible outcomes and *one* of these is a '6', so we say that the probability of throwing a six is

$$\frac{\text{the number of members of } S}{\text{the number of members of } P} = \frac{1}{6}.$$

Probability 7

(a) Suppose you wish to throw an even number with a die.

(i) The list of possible outcomes is $P = \{1, 2, 3, 4, 5, 6\}$. How many members does this set have?

(ii) List the members of the set E of even numbers on a die. How many members has it?

(iii) Write down the probability of throwing an even number.

(b) Suppose you wish to throw a number less than 3.

(i) List the members of the set P of possible outcomes. How many members has it?

(ii) List the members of the set L of numbers less than 3. How many members has it?

(iii) Write down the probability of throwing a number less than 3.

(c) Suppose you wish to throw a head with an ordinary penny.

(i) List the members of the set P of possible outcomes. How many members has it? Is it reasonable to assume that the outcomes are equally likely?

(ii) List the members of the set H in which you are interested. How many members has it?

(iii) What is the probability of throwing a head?

(d) If you toss a two-headed penny, do you think it is equally likely that you will get a head or a tail?

(i) List the members of the set P of possible outcomes. How many members has it?

(ii) What is the probability of throwing a head?

(iii) What is the probability of not throwing a head? What is the probability of throwing a tail?

(e) Suppose you wish to draw an ace from a pack of cards.

(i) How many possible outcomes are there? Are the outcomes all equally likely?

(ii) How many of the outcomes give an ace?

(iii) Write down the probability of drawing an ace.

The probability of an event happening is

$$\frac{\text{the number of outcomes which include the event}}{\text{the total number of possible outcomes}}$$

provided that the *possible outcomes are all equally likely.*

What is probability?

Exercise A

1 John tosses a die. What is the probability that he throws:
   (a) a '3';
   (b) a '5';
   (c) an odd number;
   (d) a number greater than 4?

2 Anne draws a card from a pack. What is the probability that she draws:
   (a) a king;
   (b) a seven;
   (c) a diamond;
   (d) a black card?

3 A bag contains a red ball, a blue ball and a yellow ball. Jean takes a ball from the bag without looking. The balls are all the same size and it is equally likely that she will choose any one of them. What is the probability that she takes:
   (a) the yellow ball;
   (b) the blue ball;
   (c) a ball that is not red?

4 A bag contains only orange-flavoured sweets and Peter takes one without looking. What is the probability that he takes:
   (a) an orange-flavoured sweet;
   (b) a lemon-flavoured sweet?

5 The numbers 1 to 10 inclusive are placed in a hat. Brenda takes a number without looking. What is the probability that she draws:
   (a) the number 7;
   (b) an even number;
   (c) a prime number;
   (d) a number greater than 6?

6 A football match can end in one of three ways: a home win (H), a draw (D), or an away win (A). Is it sensible to say that the probability of a home win is $\frac{1}{3}$?

2. EQUALLY LIKELY OUTCOMES

(a) A bag contains 3 red marbles and 1 blue one. Suppose you take a marble without looking.

We could say that the set of possible outcomes is {red marble, blue marble}. But the probability that you pick the blue marble is not $\frac{1}{2}$. Why not?

Probability 7

Drawing a red marble and drawing a blue marble are not equally likely outcomes since there are more red marbles than blue ones.

If we wish the set of possible outcomes to be equally likely, then we must write

$$\{P = red_1, red_2, red_3, blue\}.$$

Can you now write down the probability of picking the blue marble? What is the probability that you get a red marble?

(b) Two coins are tossed together. Peter lists the set of possible outcomes as

$$\{2 \text{ heads, a head and a tail, 2 tails}\}.$$

John decides to write H for head and T for tail and lists the outcomes as

$$\{HH, HT, TH, TT\}.$$

Has Peter made a list of equally likely outcomes? Has John?

Peter says the probability of getting 2 heads is $\frac{1}{3}$. John says it is $\frac{1}{4}$. Who is right?

When you tossed 2 coins in Experiment 2 of Chapter 4, what was your success fractions for getting 2 heads? Does it suggest that Peter is right or that John is right?

(c) Writing H for head and T for tail, copy and complete the table in Figure 1 to show the possible outcomes when 3 pennies are tossed together.

3 heads	H H H
2 heads and 1 tail	H H T H T H T H H
2 tails and 1 head	
3 tails	

Fig. 1

How many equally likely outcomes are there?
What is the probability of throwing:

    (i) 3 heads;
    (ii) 2 heads and 1 tail;
    (iii) 1 head and 2 tails;
    (iv) 3 tails?

Compare your probabilities with the success fractions which you obtained from Experiment 3 of Chapter 4.

Equally likely outcomes

(d) Suppose you have two dice, one red and one blue, and that you toss both dice together.

If the red die shows a '1', what numbers could the blue die show?

What numbers could the blue die show if the red die shows (i) a '2', (ii) a '3', (iii) a '4', (iv) a '5', (v) a '6'?

We shall write the number on the red die first and the number on the blue die second, so that, for example, the ordered pair (1, 4) means a '1' on the red die and a '4' on the blue. How many different ordered pairs are there? Is (1, 4) the same as (4, 1)?

The ordered pair (1, 4) gives a total score of 5. What possible scores can you get when you toss two dice? List these scores. Are these scores all equally likely?

Is each ordered pair equally likely?

Copy and complete the table in Figure 2 to show the possible outcomes when two dice are thrown.

Score	Ordered pairs giving this score
2	(1, 1)
3	(1, 2) (2, 1)
4	(1, 3) (2, 2) (3, 1)
5	(1, 4) (2, 3) (3, 2) (4, 1)
6	
7	
8	
9	
10	
11	
12	(6, 6)

Fig. 2

How many equally likely outcomes are there?

Find the probability of getting each of the possible scores.

Compare your probabilities with the success fractions which you obtained from Experiment 4 of Chapter 4. You may then have been surprised to find that the chance of scoring 7, say, is considerably more than that of scoring 12. You should now understand why this is so.

Probability

Instead of making a table, we can plot the ordered pairs as coordinates and show the possible outcomes on a diagram (see Figure 3).

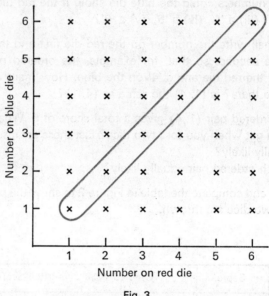

Fig. 3

How many crosses are there in the diagram? What do they represent? What do the crosses inside the red loop represent? What is the probability of throwing a double (two equal numbers)?

Copy the diagram onto squared or spotty paper. Put a ring round those crosses which represent a total greater than 8. What is the probability of throwing a total greater than 8?

Exercise B

1. A box contains 7 red biros and 3 blue ones. George takes a biro without looking. What is the probability that he
   (a) takes a red one;
   (b) takes a blue one?

2. A bag contains 40 balls. 5 are green, 15 are black and the rest are yellow. Marion takes a ball from the bag without looking. Find the probability that she takes
   (a) a black ball;
   (b) a yellow ball;
   (c) a ball which is not green.

Equally likely outcomes

3. A box contains 3 red, 2 yellow and 5 blue counters. Frances takes a counter from the box without looking. What is the probability that she

    (a) takes a blue counter;

    (b) takes a red counter?

    Compare your probabilities with the success fractions which you obtained from Experiment 6 of Chapter 4.

4. A penny and a die are tossed together. Represent the set of possible outcomes on a diagram like the one in Figure 3.
    Find the probability of getting:

    (a) a head and a '6';

    (b) a tail and an odd number;

    (c) a head and a number less than 5.

5. In a raffle for a box of chocolates 258 tickets are sold. What is the probability that you will win the box of chocolates if you have bought 3 tickets?

6. A 5p coin and a 10p coin are tossed together. Find the probability that

    (a) they will both turn up heads;

    (b) they will not both turn up heads.

7. What is the probability of drawing a picture card from a pack of playing cards (jokers excluded)?

8. Two dice are thrown together. What is the probability of throwing:

    (a) a total of 10;

    (b) a total greater than 10;

    (c) a total less than 10?

    What is the sum of these three probabilities? Give an explanation for your answer.

9. The names of Arthur, Brenda, Christine, Donald and Edward are put in a hat and one name is drawn out. What is the probability that

    (a) a girl is chosen;

    (b) a boy is chosen;

    (c) Christine is chosen;

    (d) Edward is not chosen?

Probability

10 Two representatives are to be chosen from Frank, Gwen, Helen, Ian and Joyce. The five names are put in a hat. Make a list of the ways in which two names can be drawn from the hat. What is the probability that

(a) two girls are chosen;
(b) two boys are chosen;
(c) a boy and a girl are chosen?

What is the sum of these three probabilities? Why?

11 Writing H for head and T for tail, make a list of the possible outcomes when 3 pennies are tossed together. If you have difficulty, look back at Figure 1. What is the probability of throwing:

(a) 3 heads;
(b) at least 2 heads;
(c) at least 1 head?

12 Writing H for head and T for tail, copy and complete the table in Figure 4 to show the possible outcomes when 4 pennies are tossed together.

4 heads	H H H H
3 heads and 1 tail	H H H T H H T H H T H H T H H H
2 heads and 2 tails	
1 head and 3 tails	
4 tails	

Fig. 4

How many equally likely outcomes are there?

Equally likely outcomes

What is the probability of throwing:
(a) 4 heads;
(b) 3 heads and 1 tail;
(c) 2 heads and 2 tails;
(d) 1 head and 3 tails;
(e) 4 tails?

What is the sum of these five probabilities? Why?

What is the probability of throwing at least 2 tails?

13 Two dice are made by putting the numbers 1, 2, 3, 4 on the faces of each of two regular tetrahedra.

The two dice are tossed together. Represent the set of possible outcomes on a diagram like the one in Figure 3. Find the probability of getting each of the possible scores.

14 A marble is placed at the top of a nail maze and allowed to roll down.

Figure 5 shows the number of different ways in which the marble could reach each of the positions illustrated in the figure.

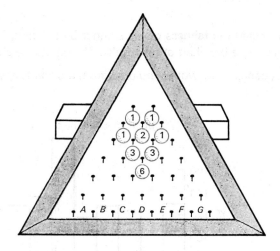

Fig. 5

By continuing the number pattern, find how many different routes lead to
(i) A, (ii) B, (iii) C, (iv) D, (v) E, (vi) F, (vii) G.

How many different routes are there altogether? If these routes are all equally likely, what is the probability that a marble will reach

(i) A, (ii) B, (iii) C, (iv) D, (v) E, (vi) F, (vii) G?

Compare your probabilities with the success fractions which you obtained from Experiment 5 of Chapter 4.

87

Probability

3. THEORY AND EXPERIMENT

(a) Charles and Stephen each used card and a cocktail stick to make an octagonal spinner like the one shown in Figure 6.

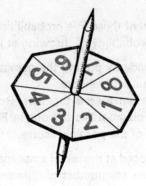

Fig. 6

Each tested the fairness of his spinner by spinning it 1000 times and and drawing a bar chart of his results. The charts are shown in Figure 7.

What conclusions would *you* draw from the charts about the fairness of the spinners?

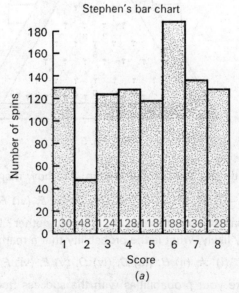

Fig. 7

Theory and experiment

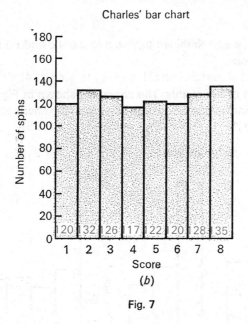

Fig. 7

Charles decided that his spinner is fair, but Stephen decided that his is biassed. Do you think that they reached sensible conclusions?

What is Charles's success fraction for scoring: (i) 2, (ii) 4, (iii) 6?

What is Stephen's success fraction for scoring: (i) 2, (ii) 4, (iii) 6?

Discuss what is meant by a 'fair' spinner.

If you use a fair spinner, how many equally likely outcomes are there? What is the probability of scoring: (i) 2, (ii) 4, (iii) 6?

Make a spinner of your own and test its fairness. Can you draw any conclusions about *your* spinner?

(*b*) In Chapter 4 you discussed ways of experimenting to find out what fraction of people who use potato peelers are left-handed. Is it reasonable to assume that being left-handed and being right-handed are equally likely? Discuss whether it is possible to find the probability that a person is left-handed by considering equally likely outcomes.

Probability

Exercise C

1 Sally, Joyce and Kathleen each made a cube and numbered its faces to form a die.

   Each tested the fairness of her die by tossing it 240 times and drawing a bar chart of her results. The charts are shown in Figure 8.

   What conclusions would you draw from the charts about the fairness of the dice?

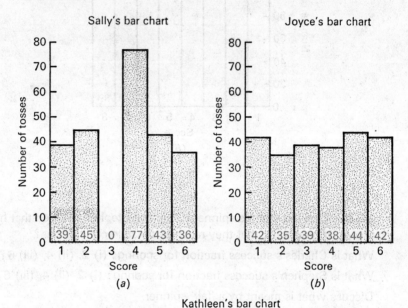

Fig. 8

Theory and experiment

2 Make a die by putting the numbers 1, 2, 3, 4 on the faces of a regular tetrahedron. Toss your die and draw a bar chart of your results. Can you draw any conclusions about the fairness of your die?

3 If you drop a drawing pin, is it just as likely to land point up as point down (see Figure 9)?

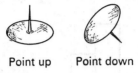

Point up   Point down

Fig. 9

Can you find the probability that a drawing-pin will land point up by considering equally likely events?

Explain how you *could* find the probability that a drawing-pin lands point up.

Do you think that the probability differs for different types of drawing pin?

4 Make a spinner like the one shown in Figure 10. Test the fairness of your spinner.

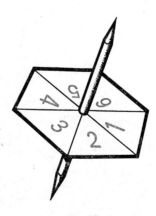

Fig. 10

Probability 7

5 John and Peter each made a nail maze like the one shown in Figure 5 on p. 87. Each tested the fairness of his maze by placing marbles at the top of the maze, letting them roll down, noting where each marble finished and recording the results in a bar chart. The charts are shown in Figure 11.

What conclusions would you draw from the charts about the fairness of the mazes?

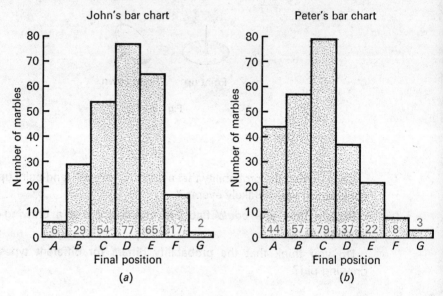

Fig. 11

8. The slide rule

---

1. MAKING AND USING A SLIDE RULE

The grids and graph papers we have met so far have all been made by marking off equal steps along both axes and this gives rise to the usual square pattern (see Figure 1).

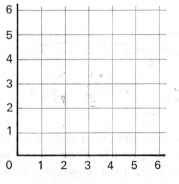

Fig. 1

The slide rule 8

Look at the piece of graph paper in Figure 2 and you will see that the axis across the page is *not* marked with equal steps.

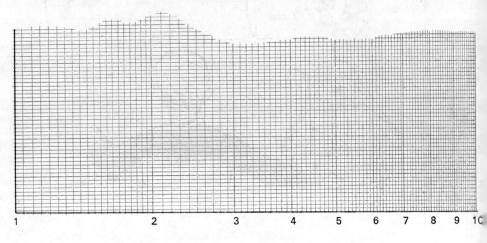

Fig. 2

On the scale going across the page, what does one small division stand for in the interval (i) between 1 and 2, (ii) between 4 and 5, (iii) between 5 and 6, (iv) between 9 and 10?

What do you notice? Why do you think this happens?

Now look at the steps in Figure 3. Where have you met this picture before?

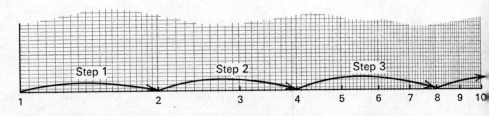

Fig. 3

You should notice that this is really the same scale as we had in the chapter on the slide rule in *Book C*, but with intermediate divisions added. With this scale it will be possible to work to a greater accuracy than before.

Making and using a slide rule

Cut two strips of this special paper and paste them carefully onto card (see Figure 4).

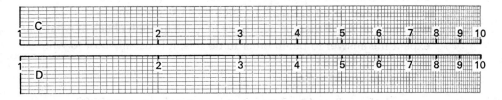

Fig. 4

Mark the major divisions, call one strip 'C' and the other 'D'.
What fact, or facts, are shown by the scales when set in the position shown in Figure 5?

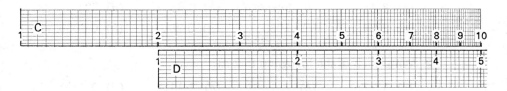

Fig. 5

Exercise A

Use your scales to calculate the following:

1. (a) 3×2; (b) $3 \times 2\frac{1}{2}$; (c) 3×3; (d) 3×3.2;
   (e) 3×1.6; (f) 3×1.65; (g) $1\frac{1}{2} \times 2\frac{1}{2}$; (h) 1.4×2.8;
   (i) 2.55×2.4; (j) 3.05×1.85; (k) 2.9×2.9; (l) $(2.35)^2$.

2. (a) $8 \div 2$; (b) $8 \div 2\frac{1}{2}$; (c) $8 \div 5$; (d) $8 \div 5.5$;
   (e) $6.4 \div 3.2$; (f) $7.65 \div 2.9$; (g) $4.05 \div 1.75$; (h) $10 \div 3$.

Which of your answers are exact? Which of your answers are only approximate?

The slide rule 8

2. ROUGH CHECKS

Did you notice that all the numbers, both in the questions and in the answers, were ten or less? In the next section you will learn how to use your slide rule to do calculations such as 41·6 × 5·3 and 620 ÷ 28. You will find that it will then be necessary to make a rough check of what you expect the answer to be. In order to do this, it is usually best to write all the numbers to 1 S.F.

Thus we would write:

41·6 × 5·3 as 40 × 5 which gives a rough answer of 200,

and   620 ÷ 28 as 600 ÷ 30 which gives a rough answer of 20.

Exercise B

1   Write each of the following to 1 S.F.

    (*a*) 7·2;   (*b*) 5·5;   (*c*) 0·8;   (*d*) 59;

    (*e*) 27·4;   (*f*) 1969;   (*g*) 0·26;   (*h*) 0·0086.

2   Write each of the following to 2 S.F.

    (*a*) 17·7;   (*b*) 231;   (*c*) 2·44;   (*d*) 0·346;

    (*e*) 1·09;   (*f*) 0·00567; (*g*) 1984;   (*h*) 10·3.

3   Write each of the following to 3 S.F.

    (*a*) 123·8;   (*b*) 22·63;   (*c*) 100·7;   (*d*) 5·892;

    (*e*) 199·6;   (*f*) 10·08;   (*g*) 1·562;   (*h*) 0·06631.

4   Write each part of Questions 2 and 3 to 1 S.F.

5   Make rough checks of the answers to each of the following by first writing each number to 1 S.F.

    (*a*) 48 × 2·7;   (*b*) 108 × 7·1;   (*c*) 63 × 67;

    (*d*) 56 ÷ 3·85;   (*e*) 68 ÷ 1·87;   (*f*) 6·8 ÷ 0·187;

    (*g*) 0·46 × 218;   (*h*) 0·061 ÷ 2·9; (*i*) 0·81 × 0·0234.

Extending the slide rule

3. EXTENDING THE SLIDE RULE

What happens when we try to set the scales for the multiplication $4 \times 5 = 20$? (See Figure 6.)

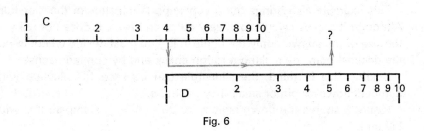

Fig. 6

Following the flow lines, the answer is off the scale! If we place another C scale immediately to the right of the one we have already, we get:

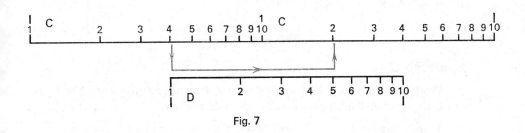

Fig. 7

The flow lines now end at 2, which in this case, really stands for 20.

However, we want to be able to do this sort of calculation with only one each of scales C and D. Remove the first C scale and alter the flow diagram, and you have:

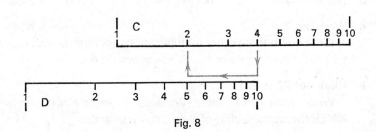

Fig. 8

The slide rule   8

> If the answer is off the scale, change ends. Instead of setting against the 1 of the D scale, set against the 10.

This example also brings out a very important feature of the slide rule. Although it makes calculations very quick and easy, it does not tell you the *size* of the answer, only the digits in it. You must decide where to put the decimal point by making a rough check and by common sense.

Bearing this in mind, the slide rule can be used to calculate with numbers however big or small they might be.

Figure 9 shows the flow diagram for 20 ÷ 5 = 4. Compare this with Figure 8.

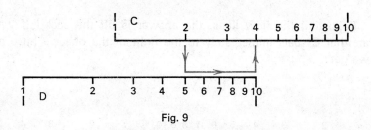

Fig. 9

Exercise C

Work out Questions 1–3 by first writing down a rough check and then using your scales to give a more accurate answer.

1. (a) 7·2 × 2·5; (b) 8·4 × 2·5; (c) 5 × 8·2;
   (d) 5·6 × 3·4; (e) 3·1 × 4·2; (f) 6·7 × 9·7;
   (g) 7·3 × 3·7; (h) (4·9)^2; (i) (8·3)^2.

2. (a) 12 ÷ 4·8; (b) 19 ÷ 2·5; (c) 17 ÷ 6·8;
   (d) 16 ÷ 4·1; (e) 25 ÷ 4·9; (f) 31 ÷ 6·6;
   (g) 34 ÷ 46; (h) 100 ÷ 11; (i) 87 ÷ 24.

3. (a) 6·1 × 72; (b) 44 × 9·1; (c) 8·2 × 33.

4. Work out 2·3 × 3·3.
   Write down three other multiplications (complete with answers) for which the same setting of the scales would do.

5. Work out 22 ÷ 3·1.
   Write down three other divisions (complete with answers) for which the same setting of the scales would do.

Accuracy

4. ACCURACY

Let us look at the multiplication 2·75 × 2·3 which we can write as

$$0 \cdot 001 \times 275 \times 23.$$

By long multiplication, or by desk calculator, we get:

$$\begin{array}{r} 275 \\ \times\ 23 \\ \hline 5500 \\ 825 \\ \hline 6325 \end{array}$$

The final answer of four figures, 6·325, is *exact*.

If we do the same multiplication on the scales, we get:

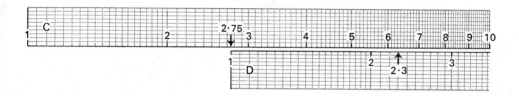

Fig. 10

If all the class did this same multiplication on their scales, the answers might be

                6·32 or 6·33 or even 6·31.

Try it and see.

How can you explain these differences? Who is right?

Suggest a sensible way of overcoming this difficulty.

Since the intermediate divisions are more widely spaced at the left-hand end of the scale than at the right, an answer appearing on the left can be read to more significant figures than an answer on the right.

The slide rule

In Figure 11 the scales are set to show 31 × 38. The exact answer is 1178.

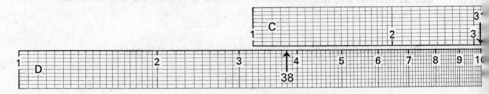

Fig. 11

What answer would you give from your scales?

> Try to give answers correct to 2 S.F. when they appear to the right of the 5 mark, and to 3 S.F. when they are to the left of the 5 mark.

Exercise D

1 Find rough answers to the following and then use the scales to work them out to either 2 S.F. or 3 S.F. as appropriate.

(a) 3·1 × 4·6; (b) 22 × 17; (c) 1·6 × 0·84;
(d) 170 × 11; (e) (8·2)²; (f) 105 × 61;
(g) ¾ × 42; (h) 430 × 31·2; (i) 4·17 × 730;
(j) 0·632 × 425; (k) 9·05 × 18·2; (l) 54·5 × 1·96;
(m) 8·1 ÷ 2·3; (n) 176 ÷ 14; (o) 2·95 ÷ 0·82;
(p) 7·08 ÷ 0·039; (q) 1970 ÷ 22; (r) 5·7 ÷ 11;
(s) 0·0047 ÷ 8·1; (t) 0·136 ÷ 8·45; (u) 2·44 ÷ 0·673.

In each of the following problems, use your scales to help you in finding an answer.

2 How many hours are there in a week?

3 How many hours are there in a year which has 365 days?

4 A car travels 17 km on 1 litre of petrol. How far would 14·5 litres take it?

Accuracy

5 Find the area of the red square in Figure 12.

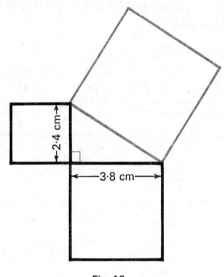

Fig. 12

6 A class of girls did a survey of leg-wear and here is a table of results:

Type of leg-wear	Number of girls having that type
Tights	13
Knee-socks	7
Ankle-socks	4
Stockings	5
Total	29

It is decided to represent this information on a pie chart; find the angle given to each girl and to each type of leg-wear.

7 Estimate the cost of 67 books each costing £0·85.

8 Find the area of a rectangle which is 4·7 cm long and 2·9 cm wide.

9 A triangle has base 17·5 cm and height 9·4 cm. Find its area.

10 £765 is to be shared between three people in the ratio 7 to 6 to 4. Find the approximate amount each one receives.

The slide rule

5. USE OF A DOUBLE SCALE

If we place two scales end to end in the length previously occupied by one scale, we get:

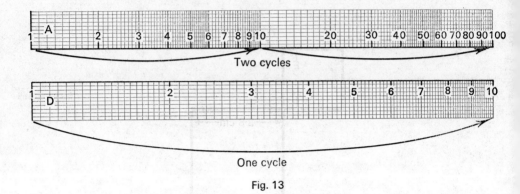

Fig. 13

Call this new scale 'A'.

What does one small division represent (i) between 1 and 2, (ii) between 6 and 7, (iii) between 10 and 20, (iv) between 60 and 70?

How do these differ from the small divisions found on scale D?

Cut a strip from each of the special 1 cycle and 2 cycle graph papers and paste them both onto pieces of card as in Figure 14. Number them as shown.

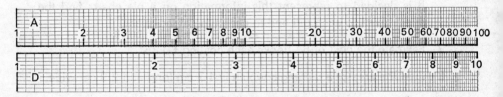

Fig. 14

Use of a double scale

Copy and complete this table:

Number on scale D	Corresponding number on scale A
1	
2	
3	
4	
5	
6	
7	
8	
9	
10	

How are the numbers on scale A related to the numbers on scale D?
How are the numbers on scale D related to the numbers on scale A?

This new arrangement of scales is going to be very useful indeed. Not only does it give squares by a direct reading without moving the scales, but, what is more important, it gives square roots.

Exercise E

Use scales A and D throughout this exercise.

1 Calculate:

(a) 17^2; (b) $(4 \cdot 5)^2$; (c) $(0 \cdot 85)^2$;

(d) $(6 \cdot 08)^2$; (e) 125^2; (f) $(0 \cdot 61)^2$;

(g) $\sqrt{20}$; (h) $\sqrt{45}$; (i) $\sqrt{6 \cdot 4}$;

(j) $\sqrt{7}$; (k) $\sqrt{0 \cdot 25}$; (l) $\sqrt{13 \cdot 5}$.

2 Find the area of a square of side 2·15 cm.

3 A square has an area of 50 cm². Find the length of a side.

4 Find the lengths of the three sides of the triangle in Figure 15 (over page).

The slide rule 8

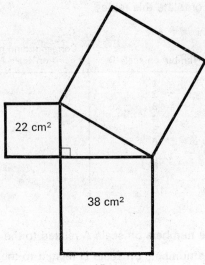

Fig. 15

6. THE SLIDE RULE AND FRACTIONS

(a) *Fractions into decimals*

Think of the fraction $\frac{3}{8}$ as $3 \div 8$ and perform this division with scales C and D. This should give the answer of 0·375.

(b) *Comparing fractions*

Make the following fractions into decimals as suggested in (a):

$$\frac{3}{5}, \frac{4}{7}, \frac{2}{3}, \frac{5}{8}.$$

Use the answers to arrange the fractions in order of size.

(c) *Equivalent fractions*

In Figure 16, scales A and B are arranged to show the set of equivalent fractions $\frac{3}{2} = \frac{6}{4} = \dots$. What other members of this set are shown on the scales?

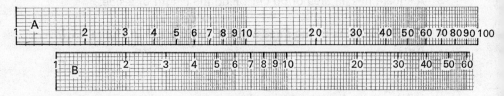

Fig. 16

The slide rule and fractions

Exercise F

1. Turn the following fractions into decimals:
   (a) $\frac{7}{8}$;   (b) $\frac{2}{7}$;   (c) $\frac{4}{9}$;   (d) $\frac{14}{5}$.

2. Use your rule to write down some fractions equivalent to:
   (a) $\frac{4}{5}$;   (b) $\frac{2}{3}$;   (c) $\frac{8}{3}$;   (d) $\frac{4}{2}$.

3. An aeroplane travels at 850 km/h for $7\frac{1}{4}$ hours. Use your rule to find the approximate distance travelled.

4. A rectangle has an area of 54 cm² and a long side of 9·4 cm. Find the length of the shorter side as accurately as your rule will allow.

5. A group of boys obtain the following marks in a test:

    9, 6, 7, 8, 3, 3, 6, 9, 8, 2, 4, 8, 8, 9, 7, 6, 5.

   Find the mean mark.

6. Find the square root of 600.

7. A machine is tested for 1000 hours' continuous running. How many days is this?

8. Triangle ABC (see Figure 17) is enlarged by a scale factor of 3·5. Write down the lengths of the sides of the enlarged figure.

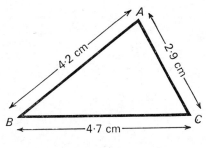

Fig. 17

9. First make the following fractions into decimals and then use your answers to make them into percentages.
   (a) $\frac{7}{20}$;   (b) $\frac{19}{25}$;   (c) $\frac{4}{15}$;   (d) $\frac{3}{8}$.

   State which answers are exact and which are approximate.

The slide rule

10 Find the length of PR in Figure 18.

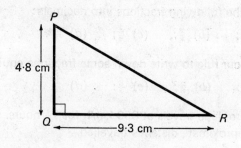

Fig. 18

11 A square has side of length 5 cm. Find the length of a diagonal.

12 What number is represented by the mark which is half-way between 1 and 2 on scale C?

13 Assuming a rate of exchange of 13·3 francs to the pound, convert the cost of the following articles advertised in a French magazine to pounds and pence:

(a) a camera—90 francs;
(b) a bottle of perfume—39 francs;
(c) a travelling case—75 francs.

Interlude

PAPER SIZES

Take two sheets of paper and fold one in half:

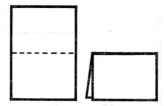

Place the folded piece in the corner of the whole sheet and obtain:

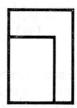

Draw a line to show whether or not the large sheet is an enlargement of the smaller sheet, as you did in the chapter on Enlargement in *Book D*.

Try many different sized pieces of paper. If you do discover some sizes which work, calculate as a decimal:

$$\frac{\text{length of long side}}{\text{length of short side}}.$$

The next section might help you to see if your answer is correct.

Interlude

International paper sizes

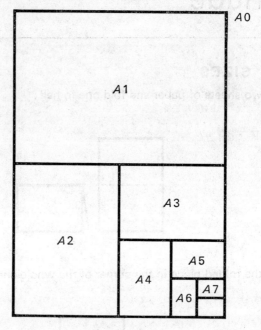

Fig. 1

The basic size, $A0$, is one square metre in area and the ratio of length to width is $\sqrt{2}$ to 1. This means that if the longer side is halved or the shorter side is doubled, the ratio of the lengths of the two sides is still $\sqrt{2}$ to 1.

This is very helpful when cutting one size from another, as there is no waste (see Figure 1), and it is particularly important with expensive papers such as photographic material.

It also leads to standardized sizes for such things as books and envelopes.

Paper sizes

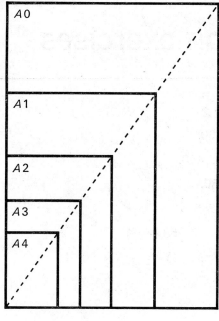

Fig. 2

In each rectangle,

$$\frac{\text{length of long side}}{\text{length of short side}} = \frac{\sqrt{2}}{1}$$

or roughly, $\frac{1\cdot 4}{1} = 1\cdot 4$.

Test to see whether your exercise books and text books are this standard shape.

Revision exercises

Computation 2
1. $25 \times \sqrt{3600}$.
2. $169708 \div 418$.
3. $20\cdot8 \times 0\cdot85$.
4. 12% of £850.
5. $3(98\cdot9 + 43\cdot7 + 53 + 6\cdot8)$.
6. $\begin{pmatrix} 2 & -1 \\ 0 & 5 \end{pmatrix} \begin{pmatrix} -3 & -1 \\ 4 & 2 \end{pmatrix}$.

Exercise G
1. Work out: (a) $(12 \div 2) \div {-3}$; (b) $12 \div (2 \div {-3})$.
   Is division associative for the set of directed numbers?
2. What is the square root of 324?
3. (a) Convert 56_{seven} to base 10; (b) convert 56_{ten} to base eight.
4. The point (5, 3) is mapped onto (2, −2) by a translation. What is the vector of the translation?
5. If $a = {-2}$ and $b = 6$, what is the value of $a(b-a)$?
6. How many planes of symmetry has a cube?
7. Divide £35 in the ratio 3 to 4.
8. What is the image set of $\{-4, -2, 0, 2, 4\}$ under the mapping $x \to x^2$? What type of correspondence is there between the two sets under this mapping?

Exercise H
1. Express $\frac{18}{40}$ as a decimal.
2. Where is the image of (4, 3) after reflection in the line $y = x$?
3. Write in shorter form: $2x + 2(4x+1)$.

Revision exercises

4 An equilateral triangle of side 5 cm is enlarged by a scale factor of $2\frac{1}{2}$. How long are the sides of the enlarged triangle?

5 Express 132 as a product of prime factors.

6 Find the value of x that fits the equation $3x-2 = 7$.

7 What is the probability of drawing an ace from a full pack of playing cards (including two jokers)?

8 Simplify $3^2 \times 3^3 \times 3^4$.

Exercise I

1 In Figure 1, $PQ = 30$ cm, $QR = 20$ cm and $\angle PQR = 90°$. Calculate the length of PR correct to 2 S.F.

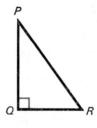

Fig. 1

2 Solve each of the following equations:
   (a) $3x-4 = 8$;
   (b) $2(x+3) = 4$;
   (c) $2x+5 = 3$;
   (d) $\frac{1}{4}(x-7) = \frac{1}{2}$.

3 In a raffle, 625 tickets are sold. What is the probability that you will win first prize if you hold 10 tickets?
   What is the probability that you will win second prize
   (a) if you do not win the first prize;
   (b) if you do win the first prize?

4 Use the C and D scales of your slide rule to calculate the following. Remember to find rough answers first.
   (a) 12.5×31;
   (b) 0.87×29;
   (c) 690×0.145;
   (d) $37 \div 0.82$;
   (e) $516 \div 9.9$;
   (f) $0.73 \div 0.046$.

Revision exercises

5 Find the sizes of the angles marked *a*, *b*, *c*, *d* in Figure 2.

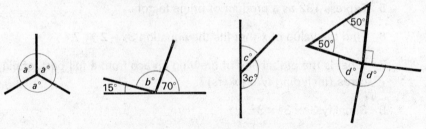

Fig. 2

6 Three authors write an article and agree to share the fee of £90 in the ratio of the approximate number of words each writes. Evans writes 2200, Francis writes 1900 and Hopkins writes 3900. How much does each get? (Since an approximate answer is required, you can use your slide rule if you like.)

Exercise J

1 Copy and complete this pattern:

$$7^2 = 49$$
$$0 \cdot 7^2 = 0 \cdot 49$$
$$0 \cdot 07^2 =$$
$$0 \cdot 007^2 =$$
$$0 \cdot 0007^2 =$$

What can you say about $\sqrt{4 \cdot 9}$; $\sqrt{0 \cdot 049}$; $\sqrt{0 \cdot 00049}$?

2 Draw the graph of $y = \frac{1}{2}x + 3$ and use it to solve the following equations:

(a) $\frac{1}{2}x + 3 = 0$; (b) $\frac{1}{2}x + 3 = 6$; (c) $\frac{1}{2}x + 3 = -2$.

3 Two dice have the numbers 1, 2, 3, 4, 5, 6 on their faces. Using ordered pairs make a list of the possible outcomes when they are thrown together. What is the probability that the two numbers which turn up have a product equal to 12?

4 Use the A and D scales of your slide rule to calculate the following. Remember to find rough answers first.

(a) 26^2; (b) $(0 \cdot 81)^2$; (c) $(4 \cdot 6)^2$; (d) $(105)^2$;
(e) $\sqrt{78}$; (f) $\sqrt{6 \cdot 8}$; (g) $\sqrt{1100}$; (h) $\sqrt{0 \cdot 067}$.

Revision exercises

5 The basic charge for hiring a Nimi car is £2 per day and in addition there is a charge of 2p per kilometre for the first 200 km and 1p per kilometre for every kilometre travelled after that. (The charge excludes the cost of the petrol.)

Copy and complete the following table for one day's motoring.

Distance travelled in km	0	50	100	200	300	400	500
Hire charge in pence	200						

Draw a graph to show the relation between the distance travelled and the hire charge.

From your graph, find the distance that would make it worthwhile paying £6·50 for one day's hire of a Nimi with unlimited use.

6 A man earning £1000 had his salary reduced by 25%. He protested and immediately had the reduced salary increased by 30%. How much did he finally gain or lose by this arrangement?

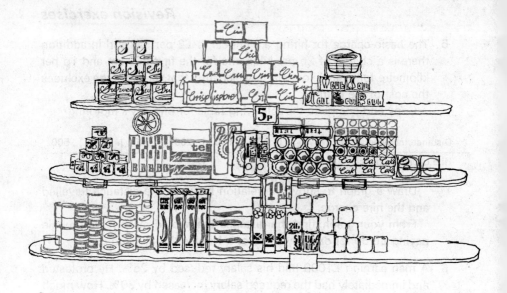

9. Volume

In the Prelude to this book the fabulous new product Sludge was mentioned. You were asked to design a packet to contain it. Some of you may have decided on a rectangular-ended box as in Figure 1. The problem would then arise as to what size to make the box in order to contain the required volume of Sludge.

Fig. 1

A reminder about area

1. A REMINDER ABOUT AREA

You will remember that in order to find the area of a flat shape it is necessary to choose a unit of area such as a triangle or a square and then see how many of these units fit into the shape. What are the areas of the shapes shown in Figure 2?

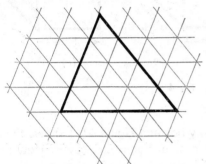
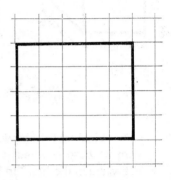

Fig. 2

Why do we prefer to use a square as the unit of area?

How many centimetre squares will fit into rectangles with the following dimensions:

(a) 4 cm by 5 cm;
(b) 8 cm by 6·5 cm?

2. UNITS OF VOLUME

In order to find the volume of a solid shape we need to choose a unit of volume. These units must be such that they will fit together to fill space without leaving gaps. You met some space fillers in the Prelude to this book. The simplest one is the cube. A cube with each edge measuring 1 cm is often used (see Figure 3). We say that its volume is 1 cubic centimetre, abbreviated to 1 cm^3. Other units of volume in the shape of a cube are the cubic metre, m^3, and the cubic millimetre, mm^3.

Fig. 3

What other possible space fillers could be used as units of volume? Why do we prefer to use a cube?

Volume

3. VOLUMES OF CUBOIDS

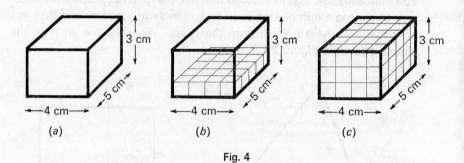

Fig. 4

(a) Figure 4 (a) shows an empty box and Figure 4 (b) shows the box with its base covered with one layer of centimetre cubes. How many cubes are in the box? In Figure 4 (c), the box has been completely filled with three layers of centimetre cubes. What is the volume of the box?

(b) Figure 5 shows a box measuring 7 cm by 3 cm by 4 cm. Copy it into your book.

Draw a sequence of diagrams as in Figure 4 and find its volume.

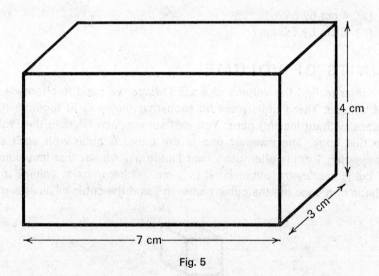

Fig. 5

(c) Without drawing, calculate the volumes of boxes with the following dimensions:

    (i) 4 cm by 3 cm by 2 cm;
    (ii) 10 cm by 12 cm by 15 cm.

Volumes of cuboids

(d) What happens when you try to fit centimetre cubes into the boxes shown in Figure 6? How would you justify the fact that the volumes can be obtained by multiplying the length by the width by the height? Find the volumes of the boxes in Figure 6.

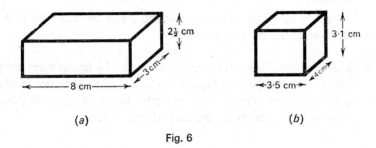

Fig. 6

Exercise A

1. Obtain a box of sugar lumps from a shop. Using one lump as a unit find the volume of the box.

2. What are the volumes of cuboids having the following measurements?

    (a) 6 cm by 3 cm by 10 cm;

    (b) $5\frac{1}{2}$ mm by 2 mm by 4 mm;

    (c) 2·5 cm by 2·5 cm by 10 cm;

    (d) 3·1 m by 6 m by 2·8 m.

3. What is the volume of a chocolate box which is 15 cm long, 8 cm wide and $4\frac{1}{2}$ cm high?

4. A long jump pit is constructed 10 m long, 2 m wide and 0·5 m deep. How many cubic metres of sand will be required to fill it?

5. I have two identical empty boxes. I fill one with tea and the other with sugar. Would you expect them to weigh the same? If not, why not?

6. 'The trench I have just dug is 5 m long, 1 m wide and 50 cm deep. I will need 250 m³ of concrete to fill it'. What is wrong with this statement?

Volume

7   A measuring container is made with a square base of edge 2 cm and vertical sides. How far apart would graduation marks on the side be if they are to indicate cubic centimetres?

8   A cuboid has a base area of 10 cm² and a height of 5 cm. What is its volume?

   Explain why the volume of a cuboid can be found by multiplying the area of the base by the height.

9   $\frac{1}{4}$ kg packets of butter measuring 11 cm by $6\frac{1}{2}$ cm by 4 cm are packed into the box shown in Figure 7. (One packet of butter is shown in position.) How many packets of butter would fill the box? What is the total weight of butter in the box when it is full?

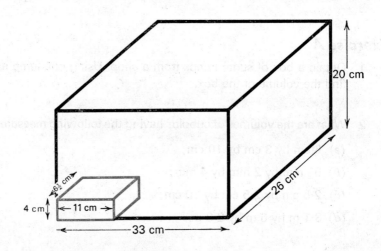

Fig. 7

10   Is it possible to completely fill a cube of side 10 cm with matchboxes measuring 5 cm by 2·5 cm by 1 cm? If so, how many will it take to fill the cube? (You might find it helpful to draw a diagram like the one in Figure 7.)

11   A builder wishes to construct a column of bricks 3 m high to support part of a house. Each layer is to be like that shown in Figure 8. If the thickness of one layer of bricks and mortar is 0·1 m, how many bricks will he require?

Volumes of cuboids

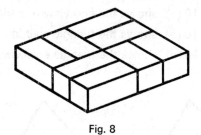

Fig. 8

12 A litre (1000 cm³) of paint covers about 10 m². What is the thickness in cm of the 'film' of paint?

Summary

The volume of a cuboid is given by

$$\text{length} \times \text{width} \times \text{height,}$$

or

$$\text{area of base} \times \text{height.}$$

4. VOLUMES OF PRISMS

Figure 9 shows a packet of Sludge in the shape of a triangular prism. The problem is to calculate its volume.

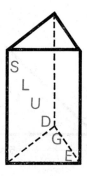

Fig. 9

Volume

(a) Figure 10 (a) shows the end view of a triangular prism.
Figure 10 (b) shows the triangle split into two parts.
Figure 10 (c) shows the triangle enclosed in a rectangle.

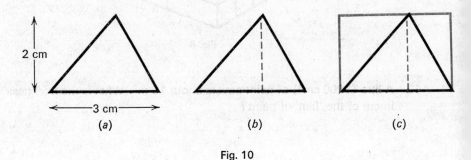

Fig. 10

(i) Explain why the area of the triangle is half the area of the rectangle.

(ii) What is the area of the rectangle?

(iii) If a prism had a rectangular end as in Figure 10 (c), and was 10 cm high, what would its volume be?

(iv) Explain why the volume of the triangular prism is 30 cm³.

(b) Figure 11 shows the end view of a triangular prism whose height is 8 cm. By using the same argument as indicated in Figure 10, find its volume.

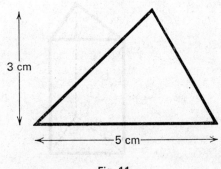

Fig. 11

(c) Explain why the volume of any triangular prism can be found by multiplying the area of the base by the height.

Volumes of prisms

(d) Figure 12 (a) shows a hexagonal packet of Sludge. Explain why its volume can be found by multiplying its base area by its height. Figure 12 (b) may help.

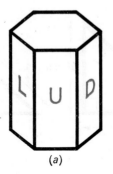
(a)

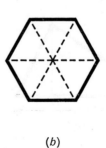
(b)

Fig. 12

(e) Figure 13 shows the end view of two more prisms. Explain why in each case their volumes can be found by multiplying their base areas by their heights.

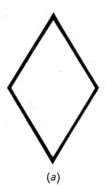
(a)

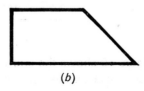
(b)

Fig. 13

Summary

The volume of any prism whose end is made up of triangles and rectangles is given by

$$\text{volume} = \text{area of base} \times \text{height}.$$

Volume

Exercise B

1. Calculate the volumes of the triangular pyramids, all of height 20 cm, whose end views are shown in Figure 14.

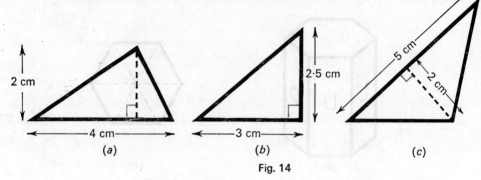

Fig. 14

2. Figure 15 shows a wedge of cheese in the shape of a triangular prism. Calculate its volume.

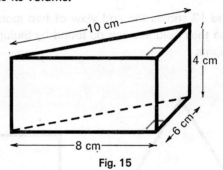

Fig. 15

3. Figure 16 shows the end view of a triangular prism of height 15 cm. Use a ruler to measure the length of the dotted line. Find the volume of the prism.

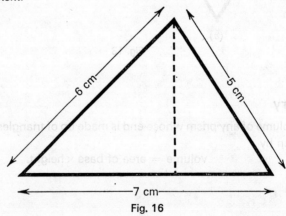

Fig. 16

Volumes of prisms

4 A prism has an end which is a triangle with sides of lengths 10 cm, 8 cm, 7 cm. Draw the triangle as accurately as you can. Find the volume of the prism if it is 12 cm high.

5 Figure 17 shows the end view of a regular hexagonal prism of height 25 cm. By splitting it up into triangles find the volume of the prism.

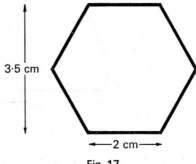

Fig. 17

6 Figure 18 shows a house built on ground 2 m above the road level. A drive 3 m wide is to be made by removing the earth indicated by the shaded region. How many cubic metres of earth will have to be moved?

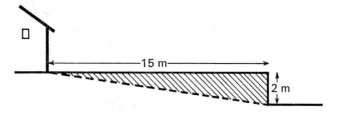

Fig. 18

7 Figure 19 shows the cross-section of a ditch which is 12 m long. Find the volume of soil which has been removed to make the ditch.

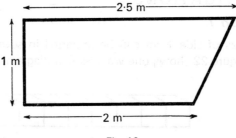

Fig. 19

Volume

8 A tent has the cross-section shown in Figure 20. If it is 2 m long, what is its volume?

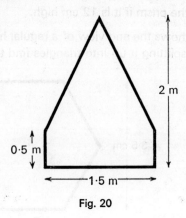

Fig. 20

9 Figure 21 shows the cross-section of a swimming pool (not drawn to scale) which is 50 m long. Find the volume of the water in the pool when it is full.

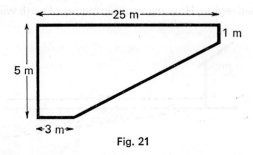

Fig. 21

10 Find the volumes of the packets you obtained for the work in the Prelude of this book.

5. INVESTIGATIONS

Investigation 1

Eight cubes of side 1 cm can be arranged in several ways to make a cuboid. Figure 22 shows one way. Sketch diagrams of the other ways.

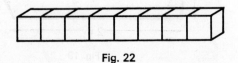

Fig. 22

124

Investigations

The amount of paper needed to cover this figure would be

$$(8 \times 4) + (1 \times 2) = 34 \text{ cm}^2.$$

Calculate the area of paper required for the other arrangements. Which shape requires the least area of paper?

Investigation 2

Repeat Investigation 1 but with 12 cubes of side 1 cm. Which is the most economical shape this time? Why does it differ from the previous investigation?

Investigation 3

Four tins have the same height but the shape of their base is different (see Figure 23). In each case the perimeter of the base is 24 cm.

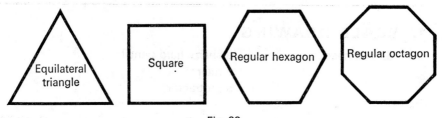

Fig. 23

Draw each figure accurately and estimate its area. Which of the tins will hold the most water?

Would you expect a cylindrical tin of circumference 24 cm and the same height as the others to hold more or less water?

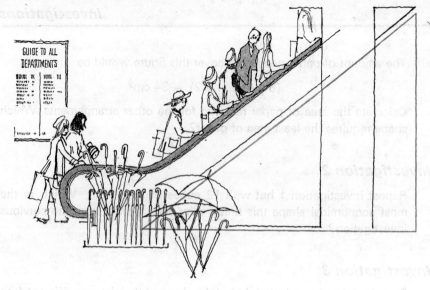

10. Enlargement

1. SCALE DRAWING

You will need   a sharp, hard pencil;
a ruler;
a protractor.

Example

In a game of hockey the ball is hit a distance of 15 m at an angle of 40° to the side line.

Figure 1 is a scale drawing showing the path of the ball from *A* to *C*.

Use your ruler to find how far the ball goes up the field, that is, the distance *AB*.

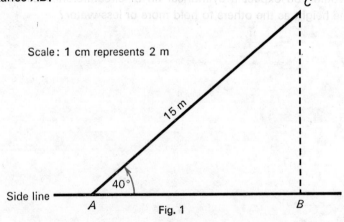

Fig. 1

Scale drawing

Exercise A

Make your drawings as large as possible.

1. A plane is climbing at an angle of 20° to the horizontal. What is its increase in height when it has gone 1000 m?

2. A 20 m ladder leans against the wall of a house so that the angle between the ladder and the ground is 50°. How far is the foot of the ladder from the wall? How high is the top of the ladder above the ground?

3. The jib of a crane is 30 m long and makes an angle of 70° with the horizontal. How high is the top of the jib above the ground?

4. An escalator in a large store is 15 m long. It makes an angle of 30° with the horizontal. What is the distance between the two floor levels?

5. A ship sails 100 km on a bearing of 050°. How far north is it from its original position? How far east?

6. A road slopes downwards at an angle of 15° to the horizontal for 300 m. What would be your loss in height in walking along it?

7. Figure 2 shows a wheel. Copy it onto graph paper making the radius 10 cm. Draw spokes at angles of 10°, 20°, 30°, 40°, 50°, 60°, 70°, 80° to the $y = 0$ axis. Read off from your diagram the coordinates of the ends of the spokes. For example, P has x-coordinate 7·7 and y-coordinate 6·4. Fill in your results in a table:

Angle	x	y
10°		
20°		
30°		
40°	7·7	6·4
50°		
60°		
70°		
80°		

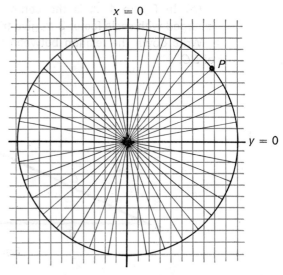

Fig. 2

Enlargement

2. ENLARGEMENT

(a) Figure 3 shows an enlargement of Mr Poly with scale factor 2.

Compare the length of his enlarged nose with the length of his original nose. How many times bigger is it?

Would the same result be true for any corresponding parts of Mr Poly?

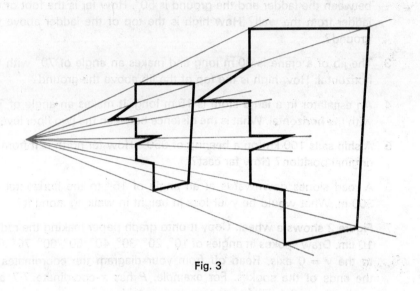

Fig. 3

(b) In Figure 4, A is the centre of enlargement. Triangle ABC is enlarged with scale factor 3 to give triangle AB'C'.

Write down the relation between the length of B'C' and the length of BC.

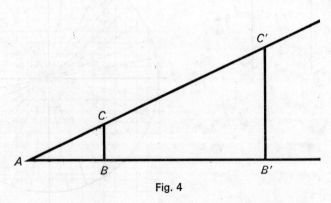

Fig. 4

Enlargement

Exercise B

1. Trace triangle *ABC* of Figure 4 and enlarge it with centre *A* and scale factors (i) 2, (ii) 3·5.

2. What is the scale factor of the enlargement in Figure 5?

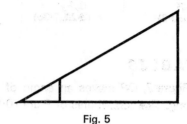

Fig. 5

3. On squared paper draw the triangle *OLM* where *O* is (0, 0), *L* is (2, 0) *M* is (2, 3).

   Enlarge it with centre *O* and scale factor 3. What are the co-ordinates of the vertices of the enlarged triangle?

   What is the relation between the new coordinates and the co-ordinates of *L* and *M*?

4. Copy the following table and fill in the coordinates of the points onto which *O*, *A* and *B* map after the following enlargements with centre *O*.

Scale factor	O(0, 0)	A(2, 0)	B(2, 1)
2			
3			
$\frac{1}{2}$			
$1\frac{1}{2}$			

5. In Figure 6, *P* maps onto *Q* under an enlargement with centre *O*. What is the scale factor of the enlargement?

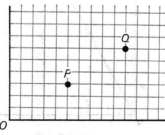

Fig. 6

Enlargement

6 *R* maps onto *S* under an enlargement with centre *O*. Copy the table and fill in the scale factors in each case.

R	S	Scale factor
(2, 1)	(4, 2)	
(3, 2)	(9, 6)	
(1, 1)	(2½, 2½)	
(1, 2)	(3½, 7)	
(2, 3)	(5, 7½)	
(1·4, 3·2)	(3·22, 7·36)	

3. USING TABLES

(a) (i) In Figure 7, *OP* makes an angle of 60° with the $y = 0$ axis and is 1 unit long. The coordinates of *P* are (0·50, 0·87).

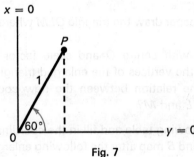

Fig. 7

(ii) In Figure 8, *OP* has been enlarged with scale factor 2. The new *y*-coordinate of *P* is $2 \times 0{\cdot}87 = 1{\cdot}74$. What is the new *x*-coordinate?

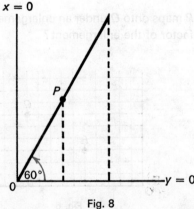

Fig. 8

(iii) In Figure 9, OP has been enlarged with scale factor 3. What are the coordinates of P'?

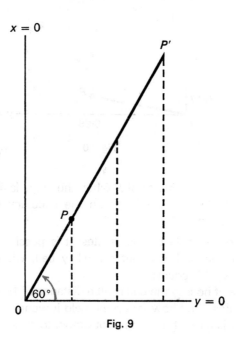

Fig. 9

(iv) If OP' was 5 units long, what would be the coordinates of P'?

You should be able to conclude from this that whatever the length of OP', it is possible to find the coordinates of P'.

(b) This table gives the coordinates of P for various angles, where OP is of length 1 unit.

Angle	x	y
10°	0·99	0·17
20°	0·94	0·34
30°	0·87	0·50
40°	0·77	0·64
50°	0·64	0·77
60°	0·50	0·87
70°	0·34	0·94
80°	0·17	0·99

Enlargement

For example, if OP is of length 1 unit at an angle of 10° to the y = 0 axis, then the coordinates of P are (0·99, 0·17) (see Figure 10).

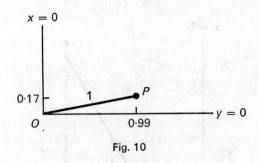

Fig. 10

What are the coordinates of P when the angle is 40°?

Use this table to check back on the accuracy of your answers to Exercise A, Question 7.

(c) Using the table of coordinates, it is possible to solve questions like the one on p. 126 about a hockey ball, without making a scale drawing. Here it is again:

In a game of hockey the ball is hit a distance of 15 m at an angle of 40° to the side line. Find how far up the field it went.

Figure 11 is a rough diagram, not drawn to scale.

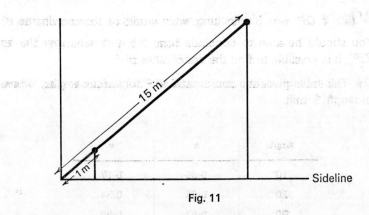

Fig. 11

If the ball had only gone a distance of 1 m then from our table we know that the distance up the field would be 0·77 m.

But it went 15 m, that is 15 times as far.

Hence the distance up the field is

$$0.77 \times 15 = 11.55 \text{ m}.$$

Using tables

Exercise C

The first five questions are reprinted from Exercise A. Use the table of coordinates to answer them, and then check back on the accuracy of your scale drawings in Exercise A.

1. A plane is climbing at an angle of 20° to the horizontal. What is its increase in height when it has gone 1000 m?

2. A 20 m ladder leans against the wall of a house so that the angle between the ladder and the ground is 50°. How far is the foot of the ladder from the wall? How high is the top of the ladder above the ground?

3. The jib of a crane is 30 m long and makes an angle of 70° with the horizontal. How high is the top of the jib above the ground?

4. An escalator in a large store is 15 m long. It makes an angle of 30° with the horizontal. What is the distance between the two floor levels?

5. A ship sails 100 km on a bearing of 050°. (What angle does its course make with the east–west line?) How far north is it from its original position? How far east?

6. Figure 12 shows the cross-section of a roof of a house. What is the width of the house?

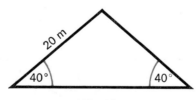

Fig. 12

7. A see-saw is 4 m long and inclined at 20° to the horizontal. How much higher is the upper end than the lower end?
   If the lower end is 0·5 m above the ground, what is the height of the upper end above the ground?

Enlargement 10

8 A boy is flying a kite on a string which is 120 m long. The string makes an angle of 60° with the ground. The boy's hand is 1·5 m above the ground (see Figure 13). What is the height of the kite above the ground?

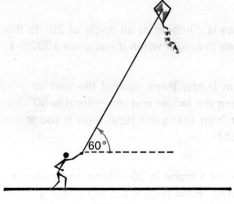

Fig. 13

If another boy is standing immediately underneath the kite, how far is he from the first boy?

9 A cliff lift is 320 m long and makes an angle of 30° with the *vertical*. How high does it rise?

10 In Figure 14, OQ is 5 units long, and the y-coordinate of Q is 2·5 units. OP is 1 unit long. What is the y-coordinate of P? By referring to the table of coordinates, find the size of angle a.

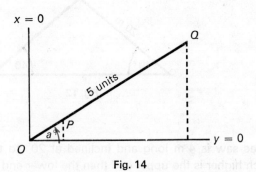

Fig. 14

Using tables

11 In Figure 15, *OS* is 4 units long and the *x*-coordinate of *S* is 2·56 units. *OR* is 1 unit long. What is the *x*-coordinate of *R*? By referring to the table of coordinates, find the size of angle *b*.

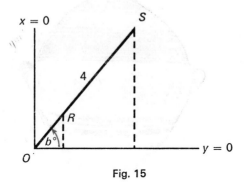

Fig. 15

12 At its steepest section, a mountain railway rises vertically through 68 m in 200 m of track length. Through what height does a train rise in travelling 1 m along the track?
At what angle is the track to the horizontal?

13 A wire rope holding a television mast is 150 m in length and is anchored to the ground 96 m from the base of the mast.
If a model was made in which the rope was 1 m long, how far would the rope have to be anchored from the base of the mast?
What is the angle between the rope and the ground?

11. The circle

1. WHAT DO WE MEAN BY A CIRCLE?

The circle has already been mentioned many times in this course, and we are now going to study the shape a little more closely.

We often hear about the circle in everyday speech;

> 'The astronauts circled the moon'.
> 'My head is going round in circles'.
> 'Events have come full circle'.

See if you can think of any other phrases which involve the circle.

In speaking of the circle, we must be clear what we mean, for there is the perfect circle of the mathematician and the approximate circle of everyday speech.

The circle is one of the shapes which early man must have noticed. Make a list of as many things as you can which occur in nature and might be called circular, and which could have been seen by our primitive ancestors. It would be best to try to list the shapes under two headings, perfect circles and approximate circles.

What do we mean by a circle?

How can you describe a circle to somebody who does not know any mathematical or technical terms? If you found yourself in the Kalahari desert without your set of mathematical drawing instruments and without a penny in your pocket, how could you show a Bushman what an approximate circle was? How could you show him what a perfect circle was?

Fig. 1

The method of drawing a circle with a piece of string and two sticks (see Figure 2), gives us a way of saying what a perfect circle is. *A circle is the set of all points in a plane at a fixed distance from a fixed point.*

Why must the string be kept tight?

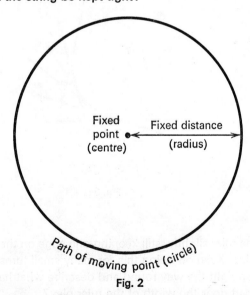

Fig. 2

When you draw a perfect circle with a pair of compasses, which is the moving point? Which is the fixed point? Which is the fixed distance?

The circle   11

Exercise A

1. (a) Make a list of as many common circular objects as you can, keeping to perfect circles this time.
   (b) Discuss why the articles you have listed are circular. How many could just as well be other shapes?

2. Draw a perfect circle using just a drawing pin, a piece of card and a pencil. Describe your method.

3. Mark a cross near the centre of a piece of paper and place a ruler against it (see Figure 3). Draw along the edge of the ruler opposite the cross.

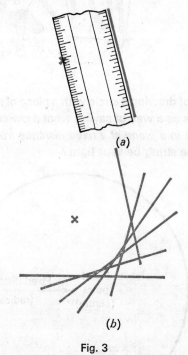

Fig. 3

Turn the ruler slightly, still keeping one edge on the cross and draw another line. Keep on doing this for many small turns of the ruler.
Carry on right the way round and describe what happens.
What part does the width of the ruler play?

4. See if you can discover other ways of drawing a circle.

2. TECHNICAL TERMS

Some of these you will have met before, but they are listed here for the sake of completeness.

Circumference The distance right round the circle; a special name for the perimeter of a circle.

Arc A part of a curved line.

Radius The distance from the centre to the circumference. (The plural of radius is radii.)

Diameter The greatest width of the circle; the distance from side to side passing through the centre. (How is this related to the radius?)

Chord Any straight line which cuts the circle into two parts. (Is the diameter a chord?)

Segment A piece of the circle cut off by a chord. (What do we call the pieces cut by the diameter?)

Sector A wedge of the circle cut out by two radii.

The circle

Exercise B

1. Draw a circle with a pair of compasses, radius 5 cm, and cut it out. Fold it in half. What do you notice about the fold line?

2. Is a chord of a circle a line of symmetry?

3. How many lines of symmetry has a circle?

4. Does a circle have rotational symmetry? If so, what is the order?

5. Place a circular shape such as a plate or a tin on a sheet of paper, draw round and cut it out. How can you find the centre of the circle?

6. Draw a circle and cut it out. Mark any two points on the circumference and fold one point over onto the other. What do you notice?

7. How could you find the centre of a circle which you could not cut out and fold?

8. Cut a circle from cartridge paper and remove a thin sector. Try to make a cone with the remaining shape. What happens as you alter the size of the sector removed?

3. MEASURING THE CIRCUMFERENCE

Suggest as many methods as you can to measure the length of the circumference and discuss which of these is likely to be the most accurate.

Draw a set of circles with the radii doubling up each time; start with radius 1 cm, then 2 cm, then 4 cm and so on. Measure the circumference of each by the method you have decided on. What happens to the circumference of a circle when you double its radius?

A circle of radius 5 cm has a circumference of 31 cm. What will be the circumference of a circle of radius: (i) 10 cm, (ii) 15 cm, (iii) 25 cm?

A class project

We shall now try to discover whether a relation exists between the length of the diameter and the circumference.

The circle in Figure 4 has a diameter of 5 cm and a measured circumference of nearly 16 cm.

How are the numbers 5 and 16 related?

Measuring the circumference

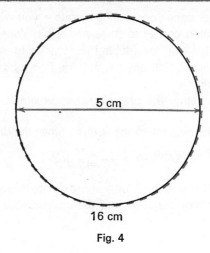

Fig. 4

You can probably think of many ways in which these two numbers are related, so we shall have to examine many circles and see if there is one relation which fits them all.

To do this, work in groups of two or three, making sure that each group draws and measures the circumference of a different sized circle. Copy the following table and enter everybody's results. Calculate the last column to 1 decimal place only.

Diameter, d (cm)	Circumference, C (cm)	$C \div d$
2		
3		
4		
5	16	3·2
6		
7		
8		
9		
10		
11		
12		

What do you notice about the last column? Can you now say how the length of the diameter is related to the circumference?

The circle

If you have all done your work accurately, you will find that each entry in the last column comes to three and a bit. Your results are bound to differ slightly due to unavoidable inaccuracies in measuring the circumference. This means that any relation we find by the above method can only be approximate.

For the time being, the best *simple* approximate relation would seem to be:

'the circumference is three times the diameter'.

For short we write this as $'C = 3d'$.

The fact that the relation is only approximate means that if we know the value of d, we can use it to find the approximate value of C and vice versa.

Exercise C

Use the approximate relation $C = 3d$ throughout.

1. Find the circumference of each of the following circles:
   - (a) diameter 6 cm;
   - (b) diameter 7 m;
   - (c) radius 11 cm;
   - (d) diameter 2·5 cm.

2. Find the circumference of:
   - (a) a plate, radius 14 cm;
   - (b) the big wheel at a fair, radius 8 m;
   - (c) a circular layout on a model railway track, diameter 4 m;
   - (d) the circle traced out by a conker whirled on a 90 cm string.

3. Assuming your neck is circular, measure its circumference and calculate its approximate diameter. Check this by another method.

4. Measure the diameter of a bicycle wheel. How far forward does the cycle travel with one revolution of its wheels?

5. Find the approximate radius of:
   - (a) an end of a cylindrical can 24 cm in circumference;
   - (b) a steering wheel 126 cm in circumference;
   - (c) a telegraph pole 75 cm in circumference;
   - (d) the trunk of a tree of 6 m girth (circumference);
   - (e) the circular wall of a city which is 5 km round.

6. Take the radius of the earth as 6400 km and calculate how far it is round the equator.

7. What length of wire would be needed to put 3 strands all round a circular cattle pen of diameter 24 m?

Measuring the circumference

8 How far does the tip of the hour hand of a town-hall clock travel each day if the hand is 1·8 m long?

9 How far does the tip of the hour hand of the school clock travel each day? How far does the tip of the minute hand travel each day?

10 Figure 5 shows a circular cycle track which has an inside radius of 100 m and is 8 m wide. How much further does a cyclist go if he keeps to the outside edge of the track rather than the inside?

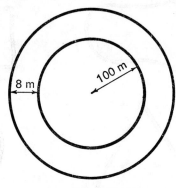

Fig. 5

11 Figure 6 shows a running track. The ends are semi-circles of radius 50 m. How long are the straights if one complete lap is 500 m?

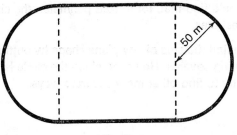

Fig. 6

12 The two circles in Figure 5 are said to be *concentric*. Find out what this means. Give some more examples of concentric circles.

The circle 11

4. THE AREA OF A CIRCLE

Take a piece of string 12 cm long, place the ends together and arrange it as a simple closed curve on a piece of squared paper. Count the squares to find the area enclosed.

Figure 7 shows some suggested shapes to try.

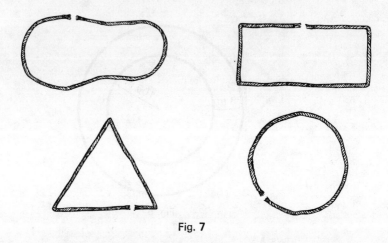

Fig. 7

Which one gives the greatest area?

You should find that, for a fixed perimeter, the circle is the shape with the greatest area.

We can find the area of any plane shape by putting it on squared paper and counting squares. However, since the circle is such a special shape, we shall try to find other more accurate ways.

The area of a circle

Method 1

Draw a square around the circle and another square inside the circle as shown in Figure 8.

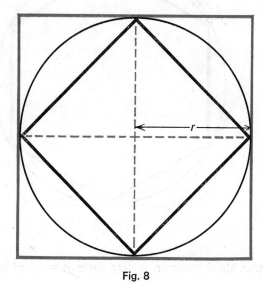

Fig. 8

Notice the eight congruent triangles in the figure. How do these show you that the area of the red square is twice the area of the black square?

If 'r' is the radius of the circle, how long is the side of the red square?

You should now be able to see that the area of the red square is $2r \times 2r = 4r^2$ square units. We found that the area of the red square is twice the area of the black square, so we now know that the area of the black square is $\frac{1}{2} \times 4r^2 = 2r^2$ square units.

Notice that the area of the circle is larger than the area of the black square and smaller than the area of the red square. That is:

area black square < area circle < area red square,

and so
$2r^2$ < area circle < $4r^2$.

If you guessed that the area of the circle was half-way between these two squares, that is, $3r^2$, you would not be far wrong. It is really 'three and a bit' times r^2, but for the time being we shall use the approximate relation:

$A = 3r^2$, where A is the area of the circle.

145

The circle

If you wanted a more accurate answer for the area of a circle you could use Figure 9.

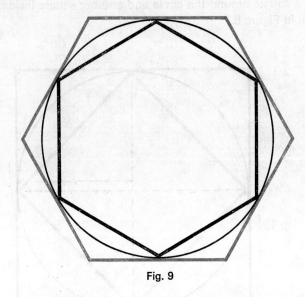

Fig. 9

area black hexagon < area circle < area red hexagon

Can you suggest what might be done to get an even more accurate answer?

Method 2

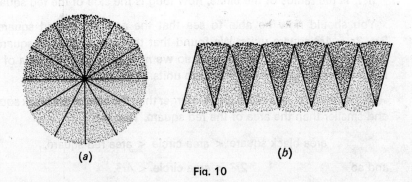

(a) (b)

Fig. 10

Study the two figures in Figure 10. Can you see what has been done to make (a) into (b)? What shape does (b) remind you of? How could you dissect the circle so that (b) became more and more like a rectangle?

The area of a circle

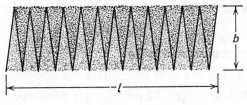

Fig. 11

If eventually you could make a rectangle this way (see Figure 11), how long would it be? How wide would it be?

You should be able to see that the length would be half the circumference and that the width would be equal to the radius of the circle.

$$\begin{aligned}\text{Area of rectangle} &= \text{length} \times \text{width} \\ &= \tfrac{1}{2} \text{ circumference} \times \text{radius} \\ &= \tfrac{1}{2} \times 6r \times r \\ &= 3r \times r \\ &= 3r^2.\end{aligned}$$

And so the area of the circle is also $3r^2$. However, as before, this is only an approximate relation. Can you say why this is so?

Exercise D

Use the approximate relation $A = 3r^2$.

1 Find the area of the circles shown in Figure 12.

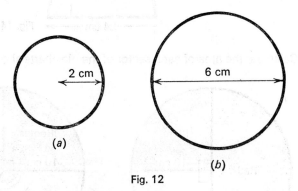

Fig. 12

2 Find the area of the circles whose radii are:
   (a) 7 mm;   (b) 4 cm;   (c) 20 m;   (d) 3·2 cm.

3 A brake disc has a diameter of 20 cm. What is its radius? What is its approximate area?

4 Find the area of a circular plate of radius 14 cm.

The circle 11

5 A circle has an area of about 75 cm². This means that, roughly,
$$3r^2 = 75.$$
Find the approximate radius.

6 Find the radii of circles which have the following areas:
(a) 300 km²;   (b) 12 m²;   (c) 150 cm².

7 A circular cattle pen has to enclose about 240 m². Obtain a rough answer for the diameter.

8 Calculate the area of the sports ground shown in Figure 13.

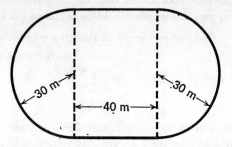

Fig. 13

9 Calculate the area of the semi-circle in Figure 14.

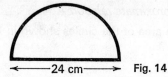

Fig. 14

10 Calculate the area of each sector of the pie charts shown in Figure 15.

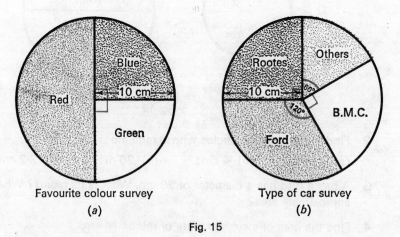

Favourite colour survey
(a)

Type of car survey
(b)

Fig. 15

The area of a circle

11 In Figure 16, calculate the area of the large circle and the area of the small circle. What is the area of the ring?

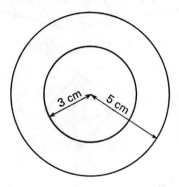

Fig. 16

12 Work out the area of each of the semi-circles in Figure 17. Add the areas of the two smallest semi-circles together. What do you notice? Does the result remind you of anything?

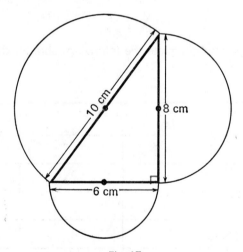

Fig. 17

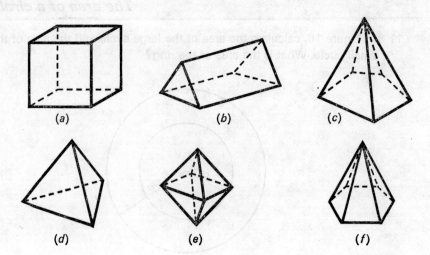

12. Networks and polyhedra

(a) Copy and complete the following table for the polyhedra shown above.

Polyhedron	Number of faces (F)	Number of vertices (V)	F+V	Number of edges (E)
(a) Cube	6			
(b) Triangular prism				9
(c) Pentagonal pyramid				
(d) Tetrahedron		4		
(e) Octahedron				
(f) Hexagonal pyramid				

Table A

What is the relation between $F+V$ and E?

150

Networks and polyhedra

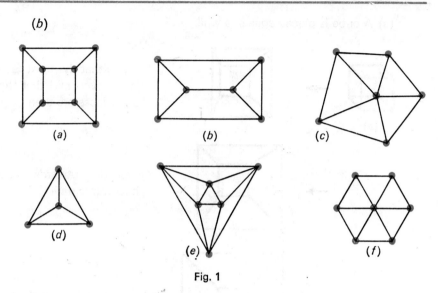

Fig. 1

Copy and complete the following table for the networks in Figure 1. Do not forget to count the outside region in each case.

Network	Number of regions (R)	Number of nodes (N)	N+R	Number of arcs (A)
(a)		8		
(b)	5			
(c)				10
(d)				
(e)				
(f)				

Table B

What is the relation between $N+R$ and A?

Compare Tables A and B. What do you notice?

Networks and polyhedra

(c) A cube is placed against a wall.

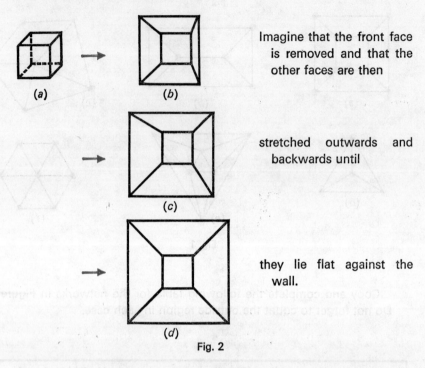

Imagine that the front face is removed and that the other faces are then stretched outwards and backwards until they lie flat against the wall.

Fig. 2

The face which was removed from the cube can be represented by the outside region of the network in Figure 2(d). This network is a topological transformation of the cube and is called a *Schlegel diagram* for the cube after a German mathematician named Schlegel.

A Schlegel diagram represents the faces, edges and vertices of a polyhedron in such a way that they can all be seen at the same time. The faces, edges and vertices of the polyhedron are equivalent to the regions, arcs and nodes of the corresponding network or Schlegel diagram.

Figure 3 shows a tetrahedron. Imagine that the front face ABC is removed and that the other faces are then stretched outwards and backwards until they lie in a plane. The result is a Schlegel diagram for a tetrahedron. Try to draw this.

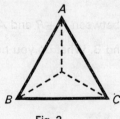

Fig. 3

Networks and polyhedra

(*d*) Draw two different Schlegel diagrams for a square-based pyramid:

    (i) by removing the square base;

    (ii) by removing one of the triangular faces.

If you have difficulty, make a skeleton model. First hold the square base towards you and then one of the triangular faces.

(*e*) The networks in Figure 1 are Schlegel diagrams for six different polyhedra. Name these polyhedra. The first one is a cube.
Why are the numbers in Table B the same as those in Table A?

(*f*) Draw a Schlegel diagram for a cube. Colour it so that regions with a common arc have different colours. Do not forget the outside region! How many colours do you need?

Make a cube from card or stiff paper. How many colours do you need to colour a cube so that faces with a common edge have different colours? (Remember that your Schlegel diagram is a topological transformation of the framework of your cube.) Colour your cube in this way using the colours yellow, blue and green. A neat and attractive result can be obtained by using coloured sticky paper.

Compare your coloured cube with those of other members of your class. Are your colour schemes the same or different? Could you form a different colour scheme from the same three colours? Make sure that faces with a common edge have different colours.

Exercise A

1 Draw a Schlegel diagram for a tetrahedron. Colour it using as few colours as possible. How many do you need?

    Make a regular tetrahedron from card or stiff paper. Colour the faces.

    Compare your model with those made by others in your class who have used the same colours. Can you colour another regular tetrahedron with the same colours so that you can tell the difference between your two tetrahedra? If your answer is 'yes', either make the model or show the colour scheme on a Schlegel diagram.

Networks and polyhedra 12

2 Draw a Schlegel diagram for a triangular prism. Show, by colouring your diagram, that four colours are needed for this polyhedron.

3 Draw a Schlegel diagram for a square-based pyramid. How many colours do you need for this polyhedron?

4 Draw a Schlegel diagram for a regular octahedron. Colour your diagram using as few colours as possible. Can you colour two regular octahedra with the same colours so that you can tell the difference between them?

5

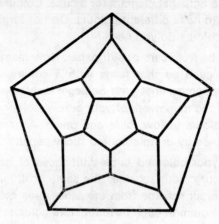

Fig. 4

Figure 4 shows a Schlegel diagram for a well-known polyhedron. How many faces has it? Which polyhedron is it?

Copy the diagram and colour it using as few colours as possible. How many do you need?

Make a model of the corresponding regular polyhedron and colour it according to the colour scheme on your diagram.

Compare your model with those of other members of your class who have used the same colours. Try to decide whether your arrangements of colours are the same or different.

Networks and polyhedra

6

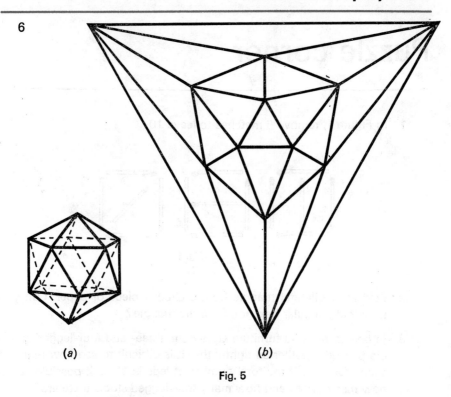

(a)　　　　　(b)

Fig. 5

Figure 5 shows a regular polyhedron and the corresponding Schlegel diagram. Which polyhedron is it? How many colours do you need for the diagram?

On four copies of the diagram, show four different arrangements of the same colours using as few as possible. There are many different arrangements for this regular polyhedron.

Puzzle corner

1. In Figure 1, remove 6 matches to leave 10.

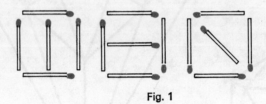

Fig. 1

2. Fold an equilateral triangle from a circular piece of paper. Can you now fold a regular hexagon from the triangle?

3. In one corner of a furniture storeroom, three- and four-legged stools are packed together so tightly that it is difficult to see how many of each there are. If the total number of legs is 41, is it possible to tell how many three- and how many four-legged stools there are?

4. Four tennis players decide on their relative merits by playing singles matches against each other. If player A beats player B, then it is accepted that A is better than B. What is the smallest number of matches which can be played to be certain of finding their order?

5. There are 2 glasses: one contains 10 spoonfuls of wine and the other 10 of water.

    A spoonful of wine is taken from the first glass, put in the second glass, and mixed round. Then a spoonful of the mixture is taken and put back in the first glass.

    Is there now more wine in the water or water in the wine?

6. Show how, with a single slice, you could cut a cube of butter so that the newly cut face would be in the shape of:

    (*a*) a triangle;   (*b*) a rectangle;   (*c*) a trapezium.

7. Trace the shaded rectangle in Figure 2. Cut out seven of these rectangles from card. Can you arrange the rectangles so that they completely cover the unshaded squares in Figure 2? How many different arrangements of the rectangles can you find? (You may find it helpful to always begin from the top left-hand corner.)

Puzzle corner

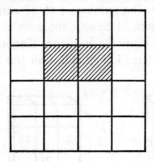

Fig. 2

8 In how many ways can the letters of the alphabet be arranged in pairs if a repetition such as AA is not allowed? (Consider AB as different from BA.)

9 Show how the shape in Figure 3 could be cut into four identical pieces.

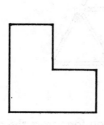 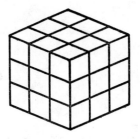

Fig. 3   Fig. 4

10 It is possible to fit 27 unit cubes together to form a cube of edge 3 units (see Figure 4). If these unit cubes have been stuck together to form the 7 shapes shown in Figure 5 they can still be assembled into the large cube. How? Try making yourself a set.

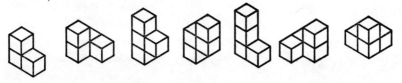

Fig. 5

Puzzle corner

11. *A magic cube.* The numbers 1 to 27 can be placed in the cells of a cube (see Figure 4) so that the sum of the numbers in every row, every column, every file and every diagonal is the same. The solution is partly completed in Figure 6. Can you finish it?

Front block　　　　Middle block　　　　Back block

Fig. 6

12. In Figure 7, remove four matches to leave two equilateral triangles. Can you leave two equilateral triangles by removing (i) three matches, (ii) just two matches?

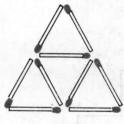

Fig. 7

13. How many different numbers can you form from the digits 1, 2, 3, 4 if each digit can only be used once in any number, but if the digits may be taken one, two, three or four at a time?

14. A barrel containing 24 litres of beer is to be shared equally by 3 men. Unfortunately, they only have available 3 containers which hold 5 litres, 11 litres and 13 litres respectively. How could they use these containers so that each man could take home 8 litres of beer?

Revision exercises

Computation 3
1. $2 \cdot 4 \times 1 \cdot 7 \times 4 \cdot 3$.
2. $(85+79+84+81+83) \div 5$.
3. $\dfrac{10 \cdot 8 \times 0 \cdot 2}{8 \times 0 \cdot 12}$.
4. $73 \cdot 78 \div 34$.
5. 4% of £85.
6. $15 \times 10 \times 5 \times 0$.

Exercise K
1. Calculate $\frac{2}{15}+\frac{1}{2}+\frac{2}{3}$.
2. If the line $y = 2x+c$ passes through the point $(1, 4)$, find the value of c.
3. What is the inverse under multiplication of:
   (a) 2; (b) $\frac{1}{3}$; (c) -4?
4. $A = \begin{pmatrix} 2 & 4 & 6 \\ 0 & -1 & -5 \end{pmatrix}$ and $B = \begin{pmatrix} 1 & 3 & 5 \\ -2 & 0 & -3 \end{pmatrix}$.

   Work out: (a) $A+B$; (b) $A-B$.
5. With which of the following shapes is it possible to form tessellations?
   (a) Rectangle; (b) isosceles triangle; (c) regular pentagon;
   (d) quadrilateral; (e) regular hexagon.
6. What linear relation do the following points satisfy?

   $(-2, -9)$, $(-1, -5)$, $(0, -1)$, $(1, 3)$, $(2, 7)$.
7. Solve the equation $3(2x-3) = 15$ by drawing a flow diagram.
8. What is the volume of a cuboid measuring 1 m by 50 cm by 25 cm?

Revision exercises

Exercise L

1. What is the size of each inside angle of a regular pentagon?
2. Write each of the following numbers correct to 2 S.F.:
   (a) 37·4;   (b) 180·8;   (c) 0·0207.
3. State the probability that a throw of a die will result in a score of 3 or more.
4. If A = {equal-sided quadrilaterals} and B = {quadrilaterals with exactly 2 lines of symmetry}, draw a quadrilateral which is a member of $A \cap B$.
5. My salary of £1200 is increased by 12%. How much do I now earn?
6. Find the median of the following numbers:
   $^-8, 2, 6, {}^-5, 0, {}^-1, {}^-4, 3.$
7. Agatha has 60 postcards which she intends sharing with her friends, Beryl and Charlotte. She keeps a quarter for herself and divides the rest among Beryl and Charlotte in the ratio 2 to 1. How many cards does each get?
8. A translation has vector $\begin{pmatrix} 2 \\ -1 \end{pmatrix}$. Onto what points would the translation map
   (a) (0, 0);   (b) (−2, 1);   (c) (−3, −1)?

Exercise M

1. Carolyn is making an open cardboard box in which her young brother can keep his building blocks. The blocks are all cubes of side 5 cm, there are 48 of them and they will only just fit in the box.
   Figure 1 is a sketch of the net for the box; $ABCD$ is a square. What are its dimensions?

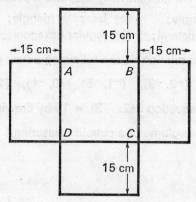

Fig. 1

Revision exercises

2. A ship sails 200 km on a bearing of 080°. Draw a sketch of this and then use the table of coordinates on p. 131 to help you calculate how far north it is from its original position. How far east or west is it from its original position?

3. A circle is cut from a piece of card with measurements as in Figure 2. Find the approximate area remaining.

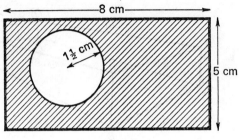

Fig. 2

4. By removing a hexagonal face each time, draw Schlegel diagrams for a hexagonal-based pyramid and a hexagonal-ended prism.

5. The handbook for a car says that the petrol tank holds 15 gallons. Use your slide rule to find the capacity of the tank in litres. Take 1 gallon as equivalent to 4·55 litres.

6. A man has a straight hedge 50 m long and he buys an electrical cutter with which to trim it. The nearest power point is 15 m away from the hedge, half-way along. Unfortunately the cable for the cutter is only 25 m long.

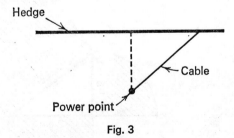

Fig. 3

Make a copy of Figure 3 and mark on it any lengths you know. Then calculate the length of the hedge the man can reach.

Use your slide rule to find by how much he should extend the cable so that he can trim the whole hedge.

Revision exercises

Exercise N

1. Figure 4 represents a swimming bath. Angles *PQR* and *QRS* are right-angles. Calculate:

   (a) the area of the side *PQRS* in square metres;
   (b) the volume of the bath in cubic metres;
   (c) how many litres of water it holds (1 litre = 1000 cm³).

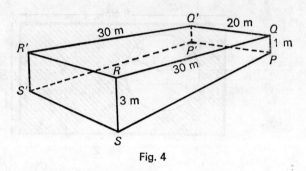

Fig. 4

2. A mountain railway in Switzerland rises steadily for 200 m at an angle of 40° to the horizontal. The bottom of the slope is 850 m above sea level. Use the table of coordinates on p. 131 to help you calculate how far above sea level the top of the slope is.

3. Find the area of each of the quadrants (¼ circles) in Figure 5. Add the two smallest areas together. What do you notice? Have you met a result like this before? Where?

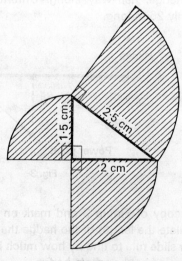

Fig. 5

Revision exercises

4 Two unbiassed spinners, numbered as shown in Figure 6, are used in a game.

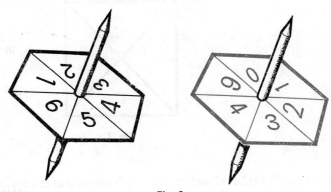

Fig. 6

Copy and complete the following table to show all the possible scores when the numbers on the two spinners are added together.

	Red spinner					
+	0	1	2	3	4	6
1						
2						
3						
4						
5						
6						

Black spinner

(a) Is one score more likely than any of the others? If so, which one?

(b) What is the probability of obtaining (i) 5, (ii) 6, (iii) 7?

(c) What is the probability of obtaining 5, 6 or 7?

(d) What is the probability that the score is a prime number?

Revision exercises

5 Sketch a solid for which Figure 7 is a Schlegel diagram.

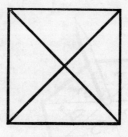

Fig. 7

6 For each of the following, say which member of the set is an 'odd man out' and why:

(a) $\{1, 4, 9, 16, 25, 36, 47, 64\}$;

(b) $\{\frac{1}{2}, \frac{3}{6}, \frac{2}{5}, \frac{9}{18}, \frac{1\frac{1}{2}}{3}\}$;

(c) $\{(0, 0), (4, 2), (3, 1), (16, 8), (9, 4\cdot5)\}$;

(d) $\{9_{ten}, 1001_{two}, 100_{three}, 13_{six}, 23_{four}\}$;

(e) {triangle, quadrilateral, tetrahedron, pentagon, hexagon}.

Exercise O

1 All 36 members of Form 3B like either porridge or cornflakes or both. If 33 like cornflakes and 24 like porridge, how many like both?

2 Pythagoras Park at Mathstown is, of course, in the shape of a right-angled triangle. The local authority decides to put street lamps at equal intervals all round the boundary. It starts by putting a lamp at each corner and four more along the shortest side (see Figure 8). How many more lamps are needed?

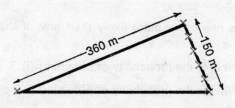

Fig. 8

Revision exercises

3   The principal means of transport to school used by a sixth form were: bus, 28%; car or motor cycle, 9%; bicycle, 39%; train, 6%; foot, 18%.

Illustrate this information on a pie chart.

4   (a) Write down the first 5 rows of Pascal's Triangle (that is, as far as the row beginning 1, 4, ...).

(b) Find the totals of the numbers in each row.

(c) Without writing down any more rows of the triangle, give the totals of the numbers in (i) the 7th row; (ii) the 10th row; (iii) the nth row.

5   ABCDE is a regular pentagon.

(a) How many lines of symmetry does the figure have?

(b) What is the order of rotational symmetry about its centre, O?

(c) What clockwise rotation about O would map A onto D?

(d) If M is the mid-point of AB, calculate the size of the angle MOC.

6   Compile the direct route matrix for the section of the London underground shown in Figure 9.

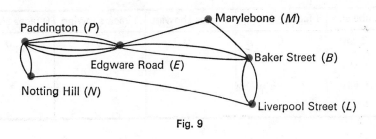

Fig. 9

Revision exercises

Exercise P

1

Distance (km)	Less than 4000	4000 to 8000	8000 to 12 000	More than 12 000
Frequency	125	257	328	90

A large tyre manufacturer kept a record of the distances after which a particular kind of tyre needed replacing. What is the probability that if you buy a tyre of this kind:

(a) it will need replacing before it has covered 4000 km;
(b) it will last more than 8000 km;
(c) it will need replacing after it has covered somewhere between 4000 and 12 000 km?

2 You can make a cube of side 2 cm with 8 cubes each of side 1 cm. Each of these cubes will have 3 of its 6 faces showing. A cube of side 3 cm needs 27 cubes of side 1 cm. How many of these will have each of the following number of faces showing?

(a) 3;   (b) 2;   (c) 1;   (d) 0.

Copy and complete the table. Comment on any patterns that you find.

Cube size	3 faces showing	2 faces showing	1 face showing	0 faces showing	Total
2 cm	8	0	0	0	8
3 cm					27
4 cm					64
5 cm					125

3 Describe how you can make a cylinder from a rectangular piece of paper.

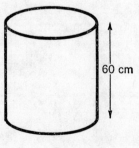

Fig. 10

Revision exercises

The tin in Figure 10 has radius 10 cm. Use the approximate relations $C = 6r$ and $A = 3r^2$ to estimate the area of material needed to cover its outside for use as a waste paper basket.

4 The operation * means 'square the first number and multiply by the second'. For example, $5 * 3 = 5^2 \times 3 = 75$.

Work out

(a) 3 * 4; (b) 4 * 3; (c) (1 * 2) * 3; (d) 1 * (2 * 3).

Is the operation (i) commutative, (ii) associative?

5 Smithson's School served 2068 dinners from Monday to Thursday inclusive.

(a) What was the average number of dinners served per day?

(b) If the daily average from Monday to Friday inclusive was 516, how many dinners were served on Friday?

(c) Dinner money was collected from every pupil at the rate of 10p per dinner. What was the total amount collected?

(d) If the full cost of the meal was 16p, what percentage of the cost was paid by the pupils?

6 A five-sided polygon *ABCDE* is symmetrical about the line $y = x$. *A* is (2, 0), *C* is (5, 5) and *D* is (3, 5). Plot these points on centimetre graph paper using 1 cm to 1 unit. What are the coordinates of *B* and *E*?

Calculate the area of the polygon.

Revision exercises

Exercise Q

Copy and complete this cross-number.

Clues Across

1. The middle term of the ninth row of Pascal's triangle.
2. The eleventh prime number.
4. How many edges has a cuboid?
6. 2^{10}.
8. Two sides of a right-angled triangle are 24 cm and 26 cm long. What is the length (in cm) of the third and shortest side?
10. $23 \cdot 1 + 124 \cdot 4 + 785 \cdot 3$ to 3 s.f.
12. 423_{five} in base ten.
14. The median of 129, 140, 128, 131, 133, 139.
15. The number of degrees in $\frac{7}{12}$ of a complete turn.
16. $4^2 + (-3)^2 + 2$.
19. 125% of £1500.
21. $\sqrt{441}$.
22. $10\frac{1}{2} \div \frac{3}{4}$.
23. The length (in cm) of an arc of a circle of circumference 96 cm subtending an angle of 60° at the centre.

Clues Down

1. The perimeter (in m) of a rectangle measuring 12·25 m by 25·25 m.
2. $81 \cdot 27 \div 0 \cdot 27$.
3. The size (in degrees) of the inside angle of a regular hexagon.
5. The sum of the prime factors of 154.
7. The volume in cm^3 of a cube of side 11 cm.
9. 14_{ten} in binary.
10. $49_{\text{eleven}} + 43_{\text{eleven}}$ in base eleven.
11. £56 is divided in the ratio 3 to 4. What is the larger share (in pounds)?
12. The number of edges of a hexagonal-based pyramid.
13. Double the sum of the first five counting numbers.
16. $0 \cdot 19 \times 1480$ to the nearest whole number.
17. The mean of 760, 763, 769, 776, 785, 791.
18. The value of $c(c-1)$ if $c = {}^{-}6$.
20. How many equilateral triangles of side 3 cm will fit inside a similar triangle of side 12 cm?